T0298017

METHODS OF DIFFERENTIAL GEOMETRY IN CLASSICAL FIELD THEORIES

k-Symplectic and k-Cosymplectic Approaches

METHODS OF DIFFERENTIAL GEOMETRY IN CLASSICAL FIELD THEORIES

k-Symplectic and k-Cosymplectic Approaches

Manuel de León
Instituto de Ciencias Matemáticas, Spain

Modesto Salgado
Universidade de Santiago de Compostela, Spain

Silvia Vilariño
Centro Universitario de la Defensa, Spain

World Scientific

NEW JERSEY • LONDON • SINGAPORE • BEIJING • SHANGHAI • HONG KONG • TAIPEI • CHENNAI • TOKYO

Published by

World Scientific Publishing Co. Pte. Ltd.

5 Toh Tuck Link, Singapore 596224

USA office: 27 Warren Street, Suite 401-402, Hackensack, NJ 07601

UK office: 57 Shelton Street, Covent Garden, London WC2H 9HE

Library of Congress Cataloging-in-Publication Data

León, Manuel de, 1953–

Methods of differential geometry in classical field theories : k-symplectic and k-cosymplectic approaches / by Manuel de León (Instituto de Ciencias Matemáticas, Spain), Modesto Salgado (Universidade de Santiago de Compostela, Spain), Silvia Vilariño (Centro Universitario de la Defensa, Spain).

pages cm

Includes bibliographical references and index.

ISBN 978-9814699754 (hardcover : alk. paper)

1. Symplectic geometry. 2. Geometry, Differential. 3. Hamiltonian operator. 4. Manifolds (Mathematics) I. Salgado, Modesto. II. Vilariño, Silvia. III. Title.

QA665.D45 2015

516.3'6--dc23

2015025680

British Library Cataloguing-in-Publication Data

A catalogue record for this book is available from the British Library.

Printed in Singapore

Introduction

As it is well known symplectic geometry is the natural arena to develop classical mechanics; indeed, a symplectic manifold is locally as a cotangent bundle T^*Q of a manifold Q, so that the canonical coordinates (q^i, p_i) can be used as coordinates for the position (q^i) and the momenta (p_i). The symplectic form is just $\omega = dq^i \wedge dp_i$, and a simple geometric tool permits to obtain the Hamiltonian vector field X_H for a Hamiltonian $H = H(q^i, p_i)$. The integral curves of X_H are just the solution of the Hamilton equations

$$\frac{dq^i}{dt} = \frac{\partial H}{\partial p_i}, \quad \frac{dp_i}{dt} = -\frac{\partial H}{\partial q^i}.$$

In classical field theory, the Hamiltonian function is of the form

$$H = H(x^\alpha, q^i, p_i^\alpha)$$

where $(x^1, \ldots, x^k) \in \mathbb{R}^k$, q^i represent the components of the fields and p_i^α are the conjugate momenta. In the Lagrangian description, the Lagrangian function is

$$L = L(x^\alpha, q^i, v_\alpha^i)$$

where now v_α^i represent the derivations of the fields with respect to the space-time variables (x^α).

At the end of the '60s and the beginning of the '70s of the past century, there are some attempts to develop a convenient geometric framework to study classical field theories. This geometric setting was the so-called multisymplectic formalism, developed in a parallel but independent way by the Polish School led by W.M. Tulczyjew (see, for instance, [Kijowski and Tulczyjew (1979); Sniatycki (1970); Tulczyjew (1974)]); H. Goldschmidt and S. Sternberg [Goldschmidt and Sternberg (1973)] and the Spanish School by P.L. Garcia and A. Pérez-Rendón [García and Pérez-Rendón (1969, 1971)].

The idea was to consider, instead of the cotangent bundle T^*Q of a manifold Q, its bundle of k-forms, $\bigwedge^k Q$. Indeed, $\bigwedge^k Q$ is equipped with a tautological k-form where its differential is just a multisymplectic form. This approach was revised, among others, by G. Martin [Martin (1988,b)] and M. Gotay *et al.* [Gotay (1990, 1991,b); Gotay, Isenberg, Marsden and Montgomery (2004); Gotay, Isenberg, Marsden (2004)] and, more recently, by F. Cantrijn *et al.* [Cantrijn, Ibort and de León (1996, 1999)] or M. Muñoz-Lecanda *et al.* [Echeverría-Enríquez, Muñoz-Lecanda and Román-Roy (1996, 1998, 1999, 2000); Echeverría-Enríquez, López, Marín-Solano, Muñoz-Lecanda and Román-Roy (2004)], among others.

An alternative approach is the so-called k-symplectic geometry, which is based on the Whitney sum of k copies of the cotangent bundle T^*Q instead of the bundle of exterior k-forms $\bigwedge^k Q$. The k-symplectic formalism is a natural generalization to field theories of the standard symplectic formalism in mechanics. This formalism was developed in a parallel way in equivalent presentations by C. Günther in [Günther (1987)], A. Awane [Awane (1992, 1994); Awane and Goze (2000)], L. K. Norris [McLean and Norris (2000); Norris (1993, 1994, 1997, 2001)] and de M. de León *et al.* [de León, Marrero and Martín de Diego (2003); de León, Méndez and Salgado (1988, 1991)]. In this sense, the k-symplectic formalism is used to give a geometric description of certain kinds of field theories: in a local description, those theories whose Lagrangian does not depend on the base coordinates, $(x^1, \dots, x^k)$ (typically, the space-time coordinates); that is, the k-symplectic formalism is only valid for Lagrangians $L(q^i, v^i_\alpha)$ and Hamiltonians $H(q^i, p^\alpha_i)$ that depend on the field coordinates q^i and on the partial derivatives of the field v^i_α, or the corresponding momenta p^α_i.

Günther's paper [Günther (1987)] gave a geometric Hamiltonian formalism for field theories. The crucial device is the introduction of a vector-valued generalization of a symplectic form called a polysymplectic form. One of the advantages of this formalism is that one only needs the tangent and cotangent bundle of a manifold to develop it. In [Munteanu, Rey and Salgado (2004)] this formalism has been revised and clarified.

Let us remark here that the polysymplectic formalism developed by I.V. Kanatchikov [Kanatchikov (1998)] and the polysymplectic formalism developed by G. Sardanashvily *et al.* [Giachetta, Mangiarotti and Sardanashvily (1997, 1999); Mangiarotti and Sardanashvily (1998); Sardanashvily (1993, 1996)], based on a vector-valued form defined on some associated fiber bundle, is a different description of classical field theories of first order from the polysymplectic (or k-symplectic) formalism proposed by C. Günther.

This book is devoted to review two of the most relevant approaches to the study of classical field theories of first order, say k-symplectic and k-cosymplectic geometry. It is structured as follows. Chapter 1 is devoted to review the fundamentals of Hamilton and Lagrangian mechanics; therefore, the Hamilton and Euler-Lagrange equations are derived on the cotangent and tangent bundles of the configurations manifold, and both are related through the Legendre transformation.

In Part 2 we develop the geometric machinery behind the classical field theories of first order when the Hamiltonian or the Lagrangian function do not depend explicitly on the space-time variables. The geometric scenario is the so-called k-symplectic geometry. Indeed, instead of considering the cotangent bundle T^*Q of a manifold Q, we take the Whitney sum of k-copies of T^*Q and investigate its geometry. This study led to the introduction of a k-symplectic structure as a family of k closed 2-forms and a distribution satisfying some compatibility relations.

k-symplectic geometry allows us to derive the Hamilton-De Donder-Weyl equations. A derivation of these equations using a variational method is also included for the sake of completeness.

This part of the book also discusses the case of Lagrangian classical theory. The key geometric structure here is the so-called tangent bundle of k^1-covelocities, which can be defined using theory of jets, or equivalently as the Whitney sum of k copies of the tangent bundle TQ of a manifold Q. This geometric bundle $TQ\oplus \overset{k)}{\dots} \oplus TQ$ led us to define a generalization of the notion of vector fields, that is, a k-vector field on Q as a section of the canonical fibration $TQ\oplus \overset{k)}{\dots} \oplus TQ \to Q$. k-vector fields will play in classical field theories the same role that vector fields on classical mechanics.

Additionally, tangent bundles of k-velocities have its own geometry, which is a natural extension of the canonical almost tangent structures on tangent bundles. Both descriptions, Hamiltonian and Lagrangian ones, can be related by an appropriate extension of the Legendre transformation.

In this part we also include a recent result on the Hamilton-Jacobi theory for classical field theories in the framework of k-symplectic geometry.

Part 3 is devoted to extend the results in Part 2 to the case of Hamiltonian and Lagrangian functions depending explicitly on the space-time variables.

The geometric setting is the so-called k-cosymplectic manifolds, which is a natural extension of cosymplectic manifold. Let us recall that cosymplectic manifolds are the odd-dimensional counterpart of symplectic manifolds.

Finally, in Part 4 we relate the k-symplectic and k-cosymplectic formalism with the multisymplectic theory.

The book ends with two appendices where the fundamentals notions on symplectic and cosymplectic manifolds are presented.

In this book, manifolds are smooth, real, paracompact, connected and $\mathcal{C}^\infty$. Maps are $\mathcal{C}^\infty$. Sum over crossed repeated indices is understood.

Contents

List of Tables

PART 1

A review of Hamiltonian and Lagrangian mechanics

Chapter 1

Hamiltonian and Lagrangian Mechanics

In this chapter we present a brief review of Hamiltonian and Lagrangian mechanics; firstly on the cotangent bundle of an arbitrary manifold Q (the Hamiltonian formalism) and then on the tangent bundle (the Lagrangian formalism). Finally, we consider the general theory on an arbitrary symplectic manifold.

In the last part of this chapter we give a review of the non-autonomous mechanics using cosymplectic structures.

A complete description of Hamiltonian and Lagrangian mechanics can be found in [Abraham and Marsden (1978); Arnold (1978, 1998); Godbillon (1969); Godstein, Poole Jr. and Safko (2001); Holm, Schmah and Stoica (2009); Holm (2008); Libermann and Marle (1987); de León and Rodrigues (1989)]. There exists an alternative description of the Lagrangian and Hamiltonian dynamics using the notion of Lagrangian submanifold, this description can be found in [Tulczyjew (1976,b)].

1.1 Hamiltonian mechanics

In this section we present a review of the Hamiltonian mechanics on the cotangent bundle of an arbitrary manifold Q. Firstly we review some results on vector spaces.

1.1.1 *Algebraic preliminaries*

By an ***exterior form*** (or simply a form) on a vector space V, we mean an alternating multilinear function on that space with values in the field of scalars. The contraction of a vector $v \in V$ and an exterior form ω on V will be denoted by $\iota_v\omega$.

Let V be a real vector space of dimension $2n$, and $\omega : V \times V \to \mathbb{R}$ a skew-symmetric bilinear form. This form allows us to define the map

$$\begin{aligned} \flat : V &\to V^* \\ v &\to \flat(v) = \iota_v \omega = \omega(v, -) . \end{aligned}$$

If ω is non-degenerate ($i.e., \omega(v, w) = 0, \ \forall w \Rightarrow v = 0$) then ω is called a ***symplectic form*** and, V is said to be a ***symplectic vector space***.

Let us observe that when ω is non-degenerate, the map $\flat$ is injective. In fact,

$$\flat(v) = 0 \Leftrightarrow \omega(v, w) = 0, \quad \forall w \in V \Leftrightarrow v = 0 .$$

In this case, since $\flat$ is an injective mapping between vector spaces of the same dimension, we deduce that it is an isomorphism. Let us observe that the matrix of $\flat$ coincides with the matrix (ω_{ij}) of ω with respect to an arbitrary basis $\{e_i\}$ of V. The inverse isomorphism will be denoted by $\sharp : V^* \to V$.

The proof of the following proposition is a direct computation.

Prop 1.1. Let (V, ω) be a symplectic vector space. Then there exists a basis (Darboux basis) $\{e_1, \ldots, e_n, u_1, \ldots, u_n\}$ of V, such that

(1) $\omega = \sum_{i=1}^{n} e^i \wedge u^i$.

(2) The isomorphisms $\flat$ and $\sharp$ associated with ω are characterized by

$$\begin{aligned} \flat(e_i) &= \ \ u^i, \flat(u_i) = -e^i, \\ \sharp(e^i) &= -u_i, \sharp(u^i) = \ \ e_i . \end{aligned}$$

1.1.2 *Canonical forms on the cotangent bundle*

Let Q be a manifold of dimension n and T^*Q the cotangent bundle of Q, with canonical projection $\pi : T^*Q \to Q$ defined by $\pi(\nu_q) = q$.

If (q^i) is a coordinate system on $U \subseteq Q$, the induced fiber coordinate system (q^i, p_i) on T^*U is defined as follows

$$q^i\left(\nu_q\right) = q^i\left(q\right) , \quad p_i\left(\nu_q\right) = \nu_q\left(\left.\frac{\partial}{\partial q^i}\right|_q\right) , \quad 1 \leq i \leq n , \tag{1.1}$$

being $\nu_q \in T^*U$.

The canonical ***Liouville*** 1-***form*** θ on T^*Q is defined by

$$\theta\left(\nu_q\right)\left(X_{\nu_q}\right) = \nu_q\left(\pi_*\left(\nu_q\right)\left(X_{\nu_q}\right)\right) \tag{1.2}$$

where $\nu_q \in T^*Q$, $X_{\nu_q} \in T_{\nu_q}(T^*Q)$ and $\pi_*(\nu_q) : T_{\nu_q}(T^*Q) \to T_qQ$ is the tangent mapping of the canonical projection $\pi : T^*Q \to Q$ at $\nu_q \in T^*_qQ$.

In canonical coordinates, the Liouville 1-form θ is given by

$$\theta = p_i\, dq^i\,. \tag{1.3}$$

The Liouville 1-form let us define the closed 2-form

$$\omega = -d\theta \tag{1.4}$$

which is non-degenerate (at each point of T^*Q), such that $(T_{\nu_q}(T^*Q), \omega(\nu_q))$ is a symplectic vector space. This 2-form is called the ***canonical symplectic form*** on the cotangent bundle. From (1.3) and (1.4) we deduce that the local expression of ω is

$$\omega = dq^i \wedge dp_i\,. \tag{1.5}$$

The manifold T^*Q with its canonical symplectic form ω is the geometrical model of the **symplectic manifolds** which will be studied in Appendix A.

For each $\nu_q \in T^*Q$, $\omega(\nu_q)$ is a bilinear form on the vector space $T_{\nu_q}(T^*Q)$, and therefore we can define a vector bundle isomorphism

$$\begin{aligned} \flat : T(T^*Q) &\to T^*(T^*Q) \\ Z_{\nu_q} &\to \flat_{\nu_q}(Z_{\nu_q}) = \iota_{Z_{\nu_q}}\omega(\nu_q) = \omega(\nu_q)(Z_{\nu_q}, -) \end{aligned}$$

with inverse $\sharp : T^*(T^*Q) \to T(T^*Q)$.

Thus we have an isomorphism of $C^\infty(T^*Q)$-modules between the corresponding spaces of sections

$$\begin{aligned} \flat :\ \mathfrak{X}(T^*Q) &\longrightarrow \bigwedge{}^1(T^*Q) \\ Z \quad &\mapsto \flat(Z) = \iota_Z\omega \end{aligned}$$

and its inverse is denoted by $\sharp : \bigwedge^1(T^*Q) \longrightarrow T(T^*Q)$.

Taking into account Proposition 1.1 (or by a direct computation) we deduce the following Lemma

Lemma 1.1. *The isomorphisms $\flat$ and $\sharp$ are locally characterized by*

$$\begin{aligned} \flat\left(\frac{\partial}{\partial q^i}\right) &= dp_i,\ \flat\left(\frac{\partial}{\partial p_i}\right) = -dq^i\,, \\ \sharp(dq^i) &= -\frac{\partial}{\partial p_i},\ \sharp(dp_i) = \frac{\partial}{\partial q^i}\,. \end{aligned} \tag{1.6}$$

1.1.3 *Hamilton equations*

Let $H : T^*Q \to \mathbb{R}$ be a function, usually called ***Hamiltonian function***. Then there exists a unique vector field $X_H \in \mathfrak{X}(T^*Q)$ such that

$$\flat(X_H) = \iota_{X_H}\omega = dH \tag{1.7}$$

or, equivalently, $X_H = \sharp(dH)$.

From (1.5) and (1.7) we deduce the local expression of X_H

$$X_H = \frac{\partial H}{\partial p_i}\frac{\partial}{\partial q^i} - \frac{\partial H}{\partial q^i}\frac{\partial}{\partial p_i}\,. \tag{1.8}$$

X_H is called the ***Hamiltonian vector field*** corresponding to the Hamiltonian function H.

From (1.8) we obtain the following theorem.

Prop 1.2. Let $c : \mathbb{R} \to T^*Q$ be a curve with local expression $c(t) = (q^i(t), p_i(t))$. Then c is an integral curve of the vector field X_H if and only if $c(t)$ is solution of the following system of differential equations.

$$\left.\frac{dq^i}{dt}\right|_t = \left.\frac{\partial H}{\partial p_i}\right|_{c(t)}, \quad \left.\frac{dp_i}{dt}\right|_t = -\left.\frac{\partial H}{\partial q^i}\right|_{c(t)}, \quad 1 \leq i \leq n \tag{1.9}$$

which are known as the **Hamilton equations of the classical mechanics**.

So equation (1.7) is considered the geometric version of Hamilton equations.

We recall that Hamilton equations can also be obtained from the Hamilton Principle, for more details see for instance [Abraham and Marsden (1978)].

1.2 Lagrangian mechanics

The Lagrangian mechanics allows us to obtain the Euler-Lagrange equations from a geometric approach. In this case we work over the tangent bundle of the configuration space. In this section we present a brief summary of the Lagrangian mechanics; a complete description can be found in [Abraham and Marsden (1978); Arnold (1978); Godbillon (1969); Godstein, Poole Jr. and Safko (2001); Holm, Schmah and Stoica (2009); Holm (2008)].

1.2.1 *Geometric preliminaries*

In this section we recall the canonical geometric ingredients on the tangent bundle, TQ, of a manifold Q, as well as other objects defined from a Lagrangian L. We denote by $\tau : TQ \to Q$ the canonical projection $\tau(v_q) = q$.

If (q^i) is a coordinate system on $U \subseteq Q$, the induced coordinate system (q^i, v^i) on $TU \subseteq TQ$ is given by

$$q^i(v_q) = q^i(q)\,, \quad v^i(v_q) = \left(dq^i\right)_q(v_q) = v_q(q^i), \quad 1 \le i \le n \tag{1.10}$$

being $v_q \in TU$.

We now recall the definition of some geometric elements which are necessary for the geometric description of the Euler-Lagrange equations.

Vertical lift of vector fields.

The structure of vector space of each fiber T_qQ of TQ allows us to define the vertical lifts of tangent vectors.

Definition 1.1. Let $X_q \in T_qQ$ be a tangent vector at the point $q \in Q$. We define the mapping

$$\begin{array}{rcl} T_qQ & \longrightarrow & T_{v_q}(TQ) \\ v_q & \to & (X_q)^{\mathrm{V}}_{v_q} = \left.\dfrac{d}{dt}\right|_{t=0}(v_q + tX_q)\,. \end{array}$$

Then, the tangent vector $(X_q)^{\mathrm{V}}_{v_q}$ is called the ***vertical lift*** of X_q to TQ at the point $v_q \in TQ$, and it is the tangent vector at $0 \in \mathbb{R}$ to the curve $\alpha(t) = v_q + tX_q \in T_qQ \subset TQ$.

In local coordinates, if $X_q = a^i \left.\dfrac{\partial}{\partial q^i}\right|_q$, then

$$(X_q)^{\mathrm{V}}_{v_q} = a^i \left.\frac{\partial}{\partial v^i}\right|_{v_q}. \tag{1.11}$$

The definition can be extended for a vector field X on Q in the obvious manner.

The Liouville vector field.

Definition 1.2. The **Liouville vector field** $\triangle$ on TQ is the infinitesimal generator of the flow given by dilatations on each fiber, it is $\Phi : (t, v_q) \in \mathbb{R} \times TQ \longrightarrow e^t\, v_q \in TQ$.

Since $\Phi_{v_q}(t) = (q^i, e^t v^i)$ we deduce that, in bundle coordinates, the Liouville vector field is given by

$$\triangle = v^i \frac{\partial}{\partial v^i}\,. \tag{1.12}$$

Canonical tangent structure on TQ.

The vertical lifts let us construct a canonical tensor field of type $(1,1)$ on TQ in the following way

Definition 1.3. A tensor field J of type $(1,1)$ on TQ is defined as follows

$$\begin{aligned} J(v_q) : T_{v_q}(TQ) &\to T_{v_q}(TQ) \\ Z_{v_q} &\to J(v_q)\left(Z_{v_q}\right) = \left(\tau_*\left(v_q\right)\left(Z_{v_q}\right)\right)^{\mathrm{V}}\left(v_q\right) \end{aligned} \tag{1.13}$$

where $Z_{v_q} \in T_{v_q}(TQ)$ and $v_q \in T_qQ$.

This tensor field is called the ***canonical tangent structure*** or ***vertical endomorphism*** of the tangent bundle TQ.

From (1.11) and (1.13) we deduce that in canonical coordinates J is given by

$$J = \frac{\partial}{\partial v^i} \otimes dq^i \ . \tag{1.14}$$

1.2.2 *Second order differential equations*

In this section we shall describe a special kind of vector fields on TQ, known as second order differential equations, semisprays and semigerbes (in French) [Grifone (1972,b); Grifone and Mehdi (1999); Szilasi (2003)]. For short, we will use the term SODEs.

Definition 1.4. Let Γ be a vector field on TQ, i.e., $\Gamma \in \mathfrak{X}(TQ)$. Γ is a SODE if and only if it is a section of the map $\tau_* : T(TQ) \to TQ$, that is

$$\tau_* \circ \Gamma = id_{TQ} \tag{1.15}$$

where id_{TQ} is the identity function on TQ and $\tau : TQ \to Q$ the canonical projection.

The tangent lift of a curve $\alpha : I \subset \mathbb{R} \to Q$ is the curve $\dot{\alpha} : I \to TQ$ where $\dot{\alpha}(t)$ is the tangent vector to the curve α. Locally if $\alpha(t) = (q^i(t))$ then $\dot{\alpha}(t) = (q^i(t), dq^i/dt)$.

A direct computation shows that the local expression of a SODE is

$$\Gamma = v^i \frac{\partial}{\partial q^i} + \Gamma^i \frac{\partial}{\partial v^i},$$

and as a consequence of this local expression one obtains that its integral curves are tangent lifts of curves on Q.

Prop 1.3. Let Γ be a vector field on TQ. Γ is a SODE if and only if its integral curves are tangent lifts of curves on Q.

Proof. Let us suppose Γ a SODE, then locally

$$\Gamma = v^i \frac{\partial}{\partial q^i} + \Gamma^i \frac{\partial}{\partial v^i}$$

where $\Gamma^i \in \mathcal{C}^\infty(TQ)$, and let $\phi(t) = (q^i(t), v^i(t))$ be an integral curve of Γ. Then

$$\begin{aligned} \frac{dq^i}{dt}\Big|_t \frac{\partial}{\partial q^i}\Big|_{\phi(t)} + \frac{dv^i}{dt}\Big|_t \frac{\partial}{\partial v^i}\Big|_{\phi(t)} &= \Gamma(\phi(t)) = \phi_*(t)\Big(\frac{d}{dt}\Big|_t\Big) \\ &= v^i(\phi(t)) \frac{\partial}{\partial q^i}\Big|_{\phi(t)} + \Gamma^i(\phi(t)) \frac{\partial}{\partial v^i}\Big|_{\phi(t)}, \end{aligned}$$

thus

$$\frac{dq^i}{dt}\Big|_t = v^i(\phi(t)) = v^i(t)\,, \quad \Gamma^i(\phi(t)) = \frac{d^2 q^i}{dt}\Big|_t$$

and we deduce that $\phi(t) = \dot{\alpha}(t)$ where $\alpha(t) = (\tau \circ \phi)(t) = (q^i(t))$, and this curve $\alpha(t)$ is a solution of the following second order differential system

$$\frac{d^2 q^i}{dt^2}\Big|_t = \Gamma^i\Big(q^i(t), \frac{dq^i}{dt}\Big|_t\Big), \quad 1 \leq i \leq n\,. \tag{1.16}$$

The converse is proved in an analogous way. □

As a consequence of (1.12) and (1.14), a SODE can be characterized using the tangent structure as follows.

Prop 1.4. A vector field X on TQ is a SODE if and only if

$$JX = \triangle \tag{1.17}$$

where $\triangle$ is the Liouville vector field and J the vertical endomorphism on TQ.

□

1.2.3 *Euler-Lagrange equations*

In this subsection we shall give a geometric description of the Euler-Lagrange equations. Note that these equations can also be obtained from a variational principle.

The Poincaré-Cartan forms on TQ.

Given a Lagrangian function, that is, a function $L\colon TQ \to \mathbb{R}$, we consider the 1-form on TQ

$$\theta_L = dL \circ J \tag{1.18}$$

that is

$$\theta_L(v_q) : T_{v_q}(TQ) \xrightarrow{J(v_q)} T_{v_q}(TQ) \xrightarrow{dL(v_q)} \mathbb{R}$$

at each point $v_q \in TQ$.

Now we define the 2-form on TQ

$$\omega_L = -\, d\theta_L\,. \tag{1.19}$$

From (1.10) and (1.14) we deduce that

$$\theta_L = \frac{\partial L}{\partial v^i}\, dq^i, \tag{1.20}$$

and from (1.19) and (1.20) we obtain that

$$\omega_L = dq^i \wedge d\Big(\frac{\partial L}{\partial v^i}\Big) = \frac{\partial^2 L}{\partial v^i \partial q^j}\, dq^i \wedge dq^j + \frac{\partial^2 L}{\partial v^i \partial v^j}\, dq^i \wedge dv^j\,. \tag{1.21}$$

This 2-form ω_L is closed, and it is non-degenerate if and only if the matrix $\left(\dfrac{\partial^2 L}{\partial v^i \partial v^j}\right)$ is non-singular; indeed the matrix of ω_L is just

$$\begin{pmatrix} \dfrac{\partial^2 L}{\partial q^j \partial v^i} - \dfrac{\partial^2 L}{\partial q^i \partial v^j} & \dfrac{\partial^2 L}{\partial v^i \partial v^j} \\ -\dfrac{\partial^2 L}{\partial v^i \partial v^j} & 0 \end{pmatrix}.$$

Definition 1.5. A Lagrangian function $L\colon TQ \to \mathbb{R}$ is said to be **regular** if the matrix $\left(\dfrac{\partial^2 L}{\partial v^i \partial v^j}\right)$ is non-singular.

When L is regular, ω_L is non-degenerate (and hence, symplectic) and thus we can consider the isomorphism

$$\begin{aligned} \flat_L\colon\ \mathfrak{X}(TQ) &\longrightarrow \textstyle\bigwedge^1 TQ \\ Z\ \ &\mapsto \flat_L(Z) = \iota_Z \omega_L \end{aligned}$$

with inverse mapping $\sharp : \bigwedge^1(TQ) \longrightarrow \mathfrak{X}(TQ)$.

Definition 1.6. Given a Lagrangian function L, we define the ***energy*** function E_L as the function

$$E_L = \Delta(L) - L\colon TQ \to \mathbb{R}.$$

From (1.12) we deduce that E_L has the local expression

$$E_L = v^i \frac{\partial L}{\partial v^i} - L. \tag{1.22}$$

We now consider the equation

$$\flat_L(X_L) = \iota_{X_L}\omega_L = dE_L\,. \tag{1.23}$$

If we write locally X_L as

$$X_L = A^i \frac{\partial}{\partial q^i} + B^i \frac{\partial}{\partial v^i}\,, \tag{1.24}$$

where $A^i, B^i \in \mathcal{C}^\infty(TQ)$ then X_L is a solution of the equation (1.23) if and only if A^i and B^i satisfy the following system of equations:

$$\begin{aligned} \left(\frac{\partial^2 L}{\partial q^i \partial v^j} - \frac{\partial^2 L}{\partial q^j \partial v^i}\right) A^j - \frac{\partial^2 L}{\partial v^i \partial v^j} B^j &= v^j \frac{\partial^2 L}{\partial q^i \partial v^j} - \frac{\partial L}{\partial q^i}\,, \\ \frac{\partial^2 L}{\partial v^i \partial v^j} A^j &= \frac{\partial^2 L}{\partial v^i \partial v^j} v^j\,. \end{aligned} \tag{1.25}$$

If the Lagrangian is regular, then $A^i = v^i$ and we have

$$\frac{\partial^2 L}{\partial q^j \partial v^i} v^j + \frac{\partial^2 L}{\partial v^i \partial v^j} B^j = \frac{\partial L}{\partial q^i}\,. \tag{1.26}$$

Therefore when L is regular there exists a unique solution X_L, and it is a SODE. Let $\dot\alpha(t) = (q^i(t), \dfrac{dq^i}{dt})$ be an integral curve of X_L where $\alpha : t \in \mathbb{R} \to \alpha(t) = (q^i(t)) \in Q$.

From (1.16) we know that

$$\left.\frac{d^2 q^i}{dt^2}\right|_t = B^i(q^j(t), \left.\frac{dq^j}{dt}\right|_t)$$

and from (1.16) and (1.26) we obtain that the curve $\alpha(t)$ satisfies the following system of equations

$$\left.\frac{\partial^2 L}{\partial q^j \partial v^i}\right|_{\dot\alpha(t)} \left.\frac{dq^j}{dt}\right|_t + \left.\frac{\partial^2 L}{\partial v^i \partial v^j}\right|_{\dot\alpha(t)} \left.\frac{d^2 q^j}{dt^2}\right|_t = \left.\frac{\partial L}{\partial q^i}\right|_{\dot\alpha(t)} \qquad 1 \le i \le n. \tag{1.27}$$

The above equations are known as the ***Euler-Lagrange equations***. Let us observe that its solutions are curves on Q.

Prop 1.5. If L is regular then the vector field X_L solution of (1.23) is a SODE, and its solutions are the solutions of the Euler-Lagrange equations.

Usually the Euler-Lagrange equations defined by L are written as

$$\left.\frac{d}{dt}\right|_t \left(\frac{\partial L}{\partial v^i} \circ \dot\alpha\right) - \frac{\partial L}{\partial q^i} \circ \alpha = 0\,, \quad 1 \le i \le n \tag{1.28}$$

whose solutions are curves $\alpha : \mathbb{R} \to Q$. Let us observe that (1.27) are just the same equations as (1.28), but written in an extended form.

Equation (1.23) is the geometric version of the Euler-Lagrange equations, which can be obtained from Hamilton's principle, see for instance [Abraham and Marsden (1978)].

1.3 Legendre transformation

The Hamiltonian and Lagrangian formulations of mechanics are related by the Legendre transformation.

Definition 1.7. Let $L\colon TQ \to \mathbb{R}$ be a Lagrangian function; then the ***Legendre transformation*** associated to L is the map

$$\begin{aligned} FL : TQ &\to T^*Q \\ v_q &\to FL(v_q) : T_qQ \to \mathbb{R} \end{aligned}$$

defined by

$$[FL(v_q)]\,(w_q) = \frac{d}{dt}\Big|_0 L\,(v_q + t\,w_q) \tag{1.29}$$

where $v_q, w_q \in TQ$.

A direct computation shows that locally

$$FL\left(q^i, v^i\right) = \left(q^i, \frac{\partial L}{\partial v^i}\right). \tag{1.30}$$

From (1.5), (1.21) and (1.30) we deduce the following relation between the canonical symplectic form and the Poincaré-Cartan 2-form.

Prop 1.6. If ω is the canonical symplectic 2-form of the cotangent bundle T^*Q and ω_L is the Poincaré-Cartan 2-form defined in (1.19) then

$$FL^*\,\omega = \omega_L\,. \tag{1.31}$$

Prop 1.7. The following statements are equivalent

(1) $L\colon TQ \to \mathbb{R}$ is a regular Lagrangian.
(2) $FL\colon TQ \to T^*Q$ is a local diffeomorphism.
(3) ω_L is a non-degenerate, and then, a symplectic form.

Proof. The Jacobian matrix of FL is

$$\begin{pmatrix} I_n & * \\ 0 & \dfrac{\partial^2 L}{\partial v^i \partial v^j} \end{pmatrix}$$

thus FL local diffeomorphism if and only if L is regular.

On the other hand, we know that ω_L is non-degenerate if and only if L is regular. □

Definition 1.8. A Lagrangian $L\colon TQ \to \mathbb{R}$ is said to be ***hyperregular*** if the Legendre transformation $FL\colon TQ \to T^*Q$ is a global diffeomorphism.

The following result connects the Hamiltonian and Lagrangian formulations.

Prop 1.8. Let $L\colon TQ \to \mathbb{R}$ be a hyperregular Lagrangian, then we define the Hamiltonian $H : T^*Q \to \mathbb{R}$ by $H \circ FL = E_L$. Therefore, we have

$$FL_*(X_L) = X_H \,. \tag{1.32}$$

Moreover, if $\alpha : \mathbb{R} \to TQ$ is an integral curve of X_L then $FL \circ \alpha$ is an integral curve of X_H.

Proof. (1.32) is a consequence of the following: *the Euler-Lagrange equation (1.23) transforms into the Hamilton equation (1.7) via the Legendre transformation*, and conversely. □

1.4 Non-autonomous Hamiltonian and Lagrangian mechanics

In this section we consider the case of time-dependent mechanics. Now we shall give a briefly review of the geometric description of the dynamical equations in this case. As in the autonomous case this description can be extended to general cosymplectic manifolds. Thus, in Appendix B we recall the notion of cosymplectic manifolds.

1.4.1 *Hamiltonian mechanics*

Let $H\colon \mathbb{R} \times T^*Q \to \mathbb{R}$ be a time-dependent Hamiltonian. If $\pi : \mathbb{R} \times T^*Q \to T^*Q$ denotes the canonical projection, we consider $\widetilde{\omega} = \pi^*\omega$ the pull-back of the canonical symplectic 2-form on T^*Q. We shall consider bundle coordinates (t, q^i, p_i) on $\mathbb{R} \times T^*Q$.

Let us take the equations

$$\iota_{E_H} dt = 1 \,, \quad \iota_{E_H} \Omega = 0 \,, \tag{1.33}$$

where $\Omega = \widetilde{\omega} + dH \wedge dt$.

A direct computation using that locally $\widetilde{\omega} = dq^i \wedge dp_i$ shows that

$$E_H = \frac{\partial}{\partial t} + \frac{\partial H}{\partial p_i}\frac{\partial}{\partial q^i} - \frac{\partial H}{\partial q^i}\frac{\partial}{\partial p_i} \,.$$

E_H is called the ***evolution vector field*** corresponding to Hamiltonian function H. Consider now an integral curve $c(s) = (t(s), q^i(s), p_i(s))$ of the evolution vector field E_H: this implies that $c(s)$ should satisfy the following system of differential equations

$$\frac{dt}{ds} = 1, \quad \frac{dq^i}{ds} = \frac{\partial H}{\partial p_i}, \quad \frac{dp_i}{ds} = -\frac{\partial H}{\partial q^i}.$$

Since $\frac{dt}{ds} = 1$ implies $t(s) = s + constant$, we deduce that

$$\frac{dq^i}{dt} = \frac{\partial H}{\partial p_i}, \quad \frac{dp_i}{dt} = -\frac{\partial H}{\partial q^i},$$

since t is an affine transformation of s, which are the ***Hamilton equations*** for a non-autonomous Hamiltonian H.

1.4.2 *Lagrangian mechanics*

Let us consider that the Lagrangian $L(t, q^i, v^i)$ is time-dependent, then L is a function $\mathbb{R} \times TQ \to \mathbb{R}$.

Let us denote also by Δ the ***canonical vector field (Liouville vector field)*** on $\mathbb{R} \times TQ$. This vector field is the infinitesimal generator of the following flow

$$\begin{aligned} \mathbb{R} \times (\mathbb{R} \times TQ) &\longrightarrow \mathbb{R} \times TQ \\ (s, (t, v_{1q}, \ldots, v_{kq})) &\longrightarrow (t, e^s v_{1q}, \ldots, e^s v_{kq}), \end{aligned}$$

and in local coordinates it has the form $\Delta = v^i \frac{\partial}{\partial v^i}$.

Now we shall characterize the vector fields on $\mathbb{R} \times TQ$ such that their integral curves are canonical prolongations of curves on Q.

Definition 1.9. Let $\alpha : \mathbb{R} \to Q$ be a curve, we define the ***first prolongation*** $\alpha^{[1]}$ of α as the map

$$\begin{aligned} \alpha^{[1]} : \mathbb{R} &\longrightarrow \mathbb{R} \times TQ \\ t &\longrightarrow (t, \dot{\alpha}(t)) \end{aligned}$$

In an obvious way we shall consider the extension of the tangent structure J to $\mathbb{R} \times TQ$ which we denote by J and it has the same local expression $J = \frac{\partial}{\partial v^i} \otimes dq^i$.

Definition 1.10. A vector field X on $\mathbb{R} \times TQ$ is said to be a ***second order differential equation*** (SODE for short) if :

$$\iota_X dt = 1, \quad J(X) = \Delta.$$

From a direct computation in local coordinates we obtain that the local expression of a SODE X is

$$X(t, q^i, v^i) = \frac{\partial}{\partial t} + v^i \frac{\partial}{\partial q^i} + X^i \frac{\partial}{\partial v^i}. \tag{1.34}$$

As in the autonomous case, one can prove the following

Prop 1.9. X is a SODE if and only if its integral curves are prolongations of curves on Q.

In fact, if $\phi : \mathbb{R} \to \mathbb{R} \times TQ$ is an integral curve of X then ϕ is the first prolongation of $\tau \circ \phi$.

The tensor J allows us to introduce the forms Θ_L and Ω_L on $\mathbb{R} \times TQ$ as follows: $\Theta_L = dL \circ J$ and $\Omega_L = -d\Theta_L$ with local expressions

$$\Theta_L = \frac{\partial L}{\partial v^i} dq^i \quad , \quad \Omega_L = dq^i \wedge d\left(\frac{\partial L}{\partial v^i}\right). \tag{1.35}$$

Let us consider the equations

$$\iota_X dt = 1 \, , \quad \iota_{X_L} \widetilde{\Omega}_L = 0 \, , \tag{1.36}$$

where $\widetilde{\Omega}_L = \Omega_L + dE_L \wedge dt$ is the ***Poincaré-Cartan 2-form***. The Lagrangian is said to be regular if $(\partial^2 L / \partial v^i \partial v^j)$ is not singular. In this case, equations (1.36) has a unique solution X.

Theorem 1.1. *Let L be a non-autonomous regular Lagrangian on $\mathbb{R} \times TQ$ and X the vector field given by (1.36). Then X is a* SODE *whose integral curves $\alpha^{[1]}(t)$ are the solutions of*

$$\frac{d}{dt}\left(\frac{\partial L}{\partial v^i} \circ \alpha^{[1]}\right) = \frac{\partial L}{\partial q^i} \circ \alpha^{[1]},$$

*which are **Euler-Lagrange equations** for L.*

Remark 1.1. The Lagrangian and Hamiltonian mechanics can be obtained from the unified Skinner-Rusk approach, [Cortés, S. Martínez and F. Cantrijn (2002)]. On the other hand, in [Muñoz-Lecanda, Román-Roy and Yániz (2001)] the authors study the non-autonomous Lagrangian invariant by a vector field. ◇

PART 2

k-symplectic formulation of classical field theories

The symplectic geometry allows us to give a geometric description of classical mechanics (see chapter 1). On the contrary, there exist several alternative models for describing geometrically first-order classical field theories. From a conceptual point of view, the simplest one is the k-symplectic formalism, which is a natural generalization to field theories of the standard symplectic formalism.

The k-symplectic formalism (also called polysymplectic formalism of C. Günther in [Günther (1987)]) is used to give a geometric description of certain kind of classical field theories: in a local description, those whose Lagrangian and Hamiltonian functions do not depend on the coordinates on the basis (that is, the space-time coordinates). Then, the k-symplectic formalism is only valid for Lagrangians and Hamiltonians that depend on the field coordinates (q^i) and on the partial derivatives of the field (v^i_α) or the corresponding momenta (p^α_i). The foundations of the k-symplectic formalism are the k-symplectic manifolds introduced by A. Awane in [Awane (1992, 1994); Awane and Goze (2000)], the k-cotangent structures introduced by M. de León *et al.* in [de León, Marrero and Martín de Diego (2003); de León, Méndez and Salgado (1988b); de León, Merino, Oubiña and Salgado (1997)] or the n-symplectic structures on the frame bundle introduced by M. McLean and L.K. Norris [McLean and Norris (2000); Norris (1993, 1994, 1997, 2001)].

In the first chapter of this part of the book, we shall introduce the notion of k-symplectic manifold using as a model the cotangent bundle of k^1-covelocities of a manifold, that is, the Whitney sum of k-copies of the cotangent bundle. Later in section 2.2.2 we shall describe the geometric equations using the k-symplectic structures. This formulation can be applied to the study of classical field theories as we shall see in chapters 4 and 6. We present these formulations and several physical examples which can be described using this approach. Finally, we establish the equivalence between the Hamiltonian and Lagrangian formulations when the Lagrangian function satisfies some regularity property. Moreover, we shall discuss the Hamilton-Jacobi equation in the k-symplectic setting (see chapter 5).

Chapter 2

k-symplectic Geometry

The k-symplectic formulation is based on the so-called k-symplectic geometry. In this chapter we introduce the k-symplectic structure which is a generalization of the notion of symplectic structure.

We first describe the geometric model of the k-symplectic manifolds, that is the cotangent bundle of k^1-covelocities and we introduce the notion of canonical geometric structures on this manifold. The formal definition of the k-symplectic manifold is given in section 2.2.

2.1 The cotangent bundle of k^1-covelocities

We denote by $(T^1_k)^*Q$ the Whitney sum with itself of k-copies of the cotangent bundle of a manifold Q of dimension n, that is,

$$(T^1_k)^*Q = T^*Q \oplus_Q \overset{k}{\cdots} \oplus_Q T^*Q \,.$$

An element ν_q of $(T^1_k)^*Q$ is a family $(\nu_{1_q}, \dots, \nu_{k_q})$ of k covectors at the same base point $q \in Q$. Thus one can consider the canonical projection

$$\begin{aligned} \pi^k : \quad & (T^1_k)^*Q && \to Q \\ & (\nu_{1_q}, \dots, \nu_{k_q}) && \mapsto \pi^k(\nu_{1_q}, \dots, \nu_{k_q}) = q \,. \end{aligned} \tag{2.1}$$

If (q^i), with $1 \leq i \leq n$, is a local coordinate system defined on an open set $U \subseteq Q$, the induced local (bundle) coordinates system (q^i, p^α_i) on $(T^1_k)^*U = (\pi^k)^{-1}(U)$ is given by

$$q^i(\nu_{1_q}, \dots, \nu_{k_q}) = q^i(q), \qquad p^\alpha_i(\nu_{1_q}, \dots, \nu_{k_q}) = \nu_{\alpha_q}\left(\left.\frac{\partial}{\partial q^i}\right|_q \right), \tag{2.2}$$

for $1 \leq \alpha \leq k$ and $1 \leq i \leq n$.

These coordinates are called the ***canonical coordinates*** on $(T^1_k)^*Q$. Thus, $(T^1_k)^*Q$ is endowed with a smooth structure of differentiable manifold of dimension $n(k+1)$.

The following diagram shows the notation which we shall use in this book:

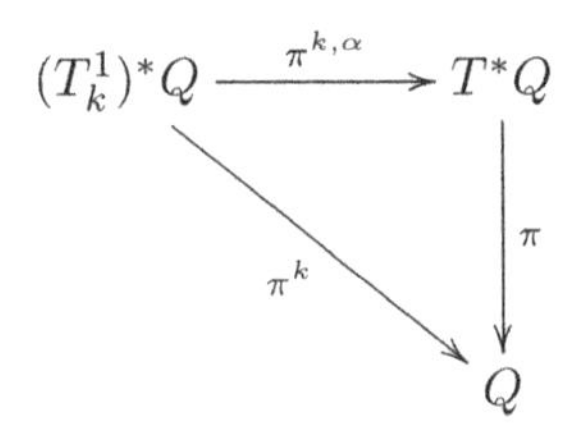

where

$$\begin{aligned} \pi^{k,\alpha}: \quad (T^1_k)^*Q &\to T^*Q \\ (\nu_{1_q},\ldots,\nu_{k_q}) &\mapsto \nu_{\alpha q} \end{aligned}, \tag{2.3}$$

is the canonical projection on each copy of the cotangent bundle T^*Q, for each $1 \le \alpha \le k$.

Remark 2.1. The manifold $(T^1_k)^*Q$ can be described using 1-jets, (we refer to [Saunders (1989)] for more details about jets).

Let $\sigma : U_q \subset Q \to \mathbb{R}^k$ and $\tau : V_q \subset Q \to \mathbb{R}^k$ be two maps defined in an open neighborhoods U_q and V_q of $q \in Q$, respectively, such that $\sigma(q) = \tau(q) = 0$. We say that σ and τ are related at $0 \in \mathbb{R}^k$ if $\sigma_*(q) = \tau_*(q)$, which means that the partial derivatives of σ and τ coincide up to order one at $q \in Q$.

The equivalence classes determined by this relationship are called *jet of order 1*, or, simply, 1-jets with source $q \in Q$ and the same target.

The 1-jet of a map $\sigma : U_q \subset Q \to \mathbb{R}^k$ is denoted by $j^1_{q,0}\sigma$ where $\sigma(q) = 0$. The set of all 1-jets at q is denoted by

$$J^1(Q,\mathbb{R}^k)_0 = \bigcup_{q\in Q} J^1_{q,0}(Q,\mathbb{R}^k) = \bigcup_{q\in Q} \{j^1_{q,0}\sigma \,|\, \sigma : Q \to \mathbb{R}^k \text{ smooth}, \sigma(q) = 0\}.$$

The canonical projection $\beta : J^1(Q,\mathbb{R}^k)_0 \to Q$ is defined by $\beta(j^1_{q,0}\sigma) = q$ and $J^1(\mathbb{R}^k,Q)_0$ is called the ***cotangent bundle of k^1-covelocities***, [Ehresmann (1951); Kolář, Michor and Slovák (1993)].

Let us observe that for $k = 1$, $J^1(Q,\mathbb{R}^k)_0$ is diffeomorphic to T^*Q.

We shall now describe the local coordinates on $J^1(\mathbb{R}^k,Q)_0$. Let U be a chart of Q with local coordinates (q^i), $1 \le i \le n$, $\sigma : U_0 \subset Q \to \mathbb{R}^k$ a mapping such that $q \in U$ and $\sigma^\alpha = x^\alpha \circ \sigma$. Then the 1-jet $j^1_{q,0}\sigma$ is uniquely represented in $\beta^{-1}(U)$ by

$$(q^i, p_i^1, \dots, p_i^k)\,, \quad 1 \leq i \leq n$$

where

$$q^i(j^1_{q,0}\sigma) = q^i(q)\,, \quad p_i^\alpha(j^1_{q,0}\sigma) = \frac{\partial \sigma^\alpha}{\partial q^i}\Big|_q = d\sigma^\alpha(q)\left(\frac{\partial}{\partial q^i}\Big|_q\right). \tag{2.4}$$

The manifolds $(T^1_k)^*Q$ and $J^1(\mathbb{R}^k, Q)_0$ can be identified, via the diffeomorphism

$$\begin{array}{ccc} J^1(\mathbb{R}^k, Q)_0 \equiv T^*Q \oplus \overset{k}{\dots} \oplus T^*Q = (T^1_k)^*Q \\ j^1_{q,0}\sigma \quad \equiv \quad (d\sigma^1(q), \dots, d\sigma^k(q)) \end{array} \tag{2.5}$$

where $\sigma^\alpha = \pi^\alpha \circ \sigma : Q \longrightarrow \mathbb{R}$ is the α-th component of σ and $\pi^\alpha \colon \mathbb{R}^k \to \mathbb{R}$ the canonical projections for each $1 \leq \alpha \leq k$. ◇

We now introduce certain canonical geometric structures on $(T^1_k)^*Q$. These structures will be used in the description of the Hamiltonian k-symplectic formalism, see section 2.2.2.

Definition 2.1. We define the ***canonical*** **1*-forms*** $\theta^1, \dots, \theta^k$ on $(T^1_k)^*Q$ as the pull-back of Liouville's 1-form θ (see (1.2)), by the canonical projection $\pi^{k,\alpha}$ (see (2.3)), that is, for each $1 \leq \alpha \leq k$

$$\theta^\alpha = (\pi^{k,\alpha})^*\theta\,;$$

the ***canonical*** **2*-forms*** $\omega^1, \dots, \omega^k$ are defined by

$$\omega^\alpha = -d\theta^\alpha$$

or equivalently by $\omega^\alpha = (\pi^{k,\alpha})^*\omega$ being ω the canonical symplectic form on the cotangent bundle T^*Q.

If we consider the canonical coordinates (q^i, p_i^α) on $(T^1_k)^*Q$ (see (2.2)), then the canonical forms $\theta^\alpha, \omega^\alpha$ have the following local expressions:

$$\theta^\alpha = p_i^\alpha dq^i\,, \quad \omega^\alpha = dq^i \wedge dp_i^\alpha\,, \tag{2.6}$$

with $1 \leq \alpha \leq k$.

Remark 2.2. An alternative definition of the canonical 1-forms $\theta^1, \dots, \theta^k$ is through the composition:

$$T_{\nu_q}((T^1_k)^*Q) \xrightarrow{(\pi^k)_*(\nu_q)} T_qQ \xrightarrow{\nu_{\alpha q}} \mathbb{R}$$

(with the composite arrow labelled $\theta^\alpha(\nu_q)$)

That is,

$$\theta^\alpha(\nu_q)(X_{\nu_q}) := \nu_{\alpha q}((\pi^k)_*(\nu_q)(X_{\nu_q})) \tag{2.7}$$

for $X_{\nu_q} \in T_{\nu_q}((T^1_k)^*Q)$, $\nu_q = (\nu_{1q}, \dots, \nu_{kq}) \in (T^1_k)^*Q$ and $q \in Q$. ◇

Let us observe that the canonical 2-forms $\omega^1,\dots,\omega^k$ are closed forms (indeed, they are exact). An interesting property of these forms is the following: for each $1\leq\alpha\leq k$, we consider the kernel of each ω^α, i.e., the set

$$\ker\omega^\alpha=\{X\in T((T^1_k)^*Q)\,|\,\iota_X\omega^\alpha=0\}\,;$$

then from (2.6) it is easy to check that

$$\omega^\alpha\Big|_{V\times V}=0 \text{ and } \bigcap_{\alpha=1}^{k}\ker\omega^\alpha=\{0\}\,, \tag{2.8}$$

where $V=\ker(\pi^k)_*$ is the vertical distribution of dimension nk associated to $\pi^k\colon(T^1_k)^*Q\to Q$. This vertical distribution is locally spanned by the set

$$\left\{\frac{\partial}{\partial p^1_1},\dots,\frac{\partial}{\partial p^k_1},\frac{\partial}{\partial p^1_2},\dots,\frac{\partial}{\partial p^k_2},\dots,\frac{\partial}{\partial p^1_n},\dots,\frac{\partial}{\partial p^k_n}\right\}\,. \tag{2.9}$$

The properties (2.8) are interesting because the family of the manifold $(T^1_k)^*Q$ with the 2-forms $\omega^1,\dots,\omega^k$ and the distribution V is the model for a k-symplectic manifold, which will be introduced in the following section.

2.2 k-symplectic geometry

A natural generalization of a symplectic manifold is the notion of the so-called k-symplectic manifold. The canonical model of a symplectic manifold is the cotangent bundle T^*Q, while the canonical model of a k-symplectic manifold is the bundle of k^1-covelocities, that is, $(T^1_k)^*Q$.

The notion of k-symplectic structure was independently introduced by A. Awane [Awane (1992); Awane and Goze (2000)], G. Günther [Günther (1987)], M. de León *et al.* [de León, Marrero and Martín de Diego (2003); de León, Méndez and Salgado (1988,b); de León, Merino, Oubiña and Salgado (1997)], and L.K. Norris [McLean and Norris (2000); Norris (1993)]. Let us recall that k-symplectic manifolds provide a natural arena to develop classical field theory as an alternative to other geometrical settings which we shall comment in the last part of this book.

A characteristic of the k-symplectic manifold is the existence of a theorem of Darboux type, therefore all k-symplectic manifolds are locally as the canonical model.

2.2.1 *k-symplectic vector spaces*

As we have mentioned above, the k-symplectic manifolds constitute the arena for the geometric study of classical field theories. This subsection considers the linear case as a preliminary step for the next subsection.

Definition 2.2. A ***k*-symplectic vector space** $(\mathcal{V}, \omega^1, \ldots, \omega^k, \mathcal{W})$ is a vector space $\mathcal{V}$ of dimension $n(k+1)$, a family of k skew-symmetric bilinear forms $\omega^1, \ldots, \omega^k$ and a vector subspace $\mathcal{W}$ of dimension nk such that

$$\bigcap_{\alpha=1}^{k} \ker \omega^\alpha = \{0\}\,, \tag{2.10}$$

where

$$\ker \omega^\alpha = \{u \in \mathcal{V} \,|\, \omega^\alpha(u, v) = 0,\ \forall v \in \mathcal{V}\}$$

denotes the kernel of ω^α and

$$\omega^\alpha\big|_{\mathcal{W}\times\mathcal{W}} = 0\,,$$

for $1 \leq \alpha \leq k$.

The condition (2.10) means that the induced linear map

$$\begin{aligned} \sharp_\omega \colon\ \mathcal{V} &\to \mathcal{V}^* \times \overset{k}{\cdots} \times \mathcal{V}^* \\ v &\mapsto (\iota_v\omega^1, \ldots, \iota_v\omega^k) \end{aligned} \tag{2.11}$$

is injective, or equivalently, that it has maximal rank, that is, $rank\, \sharp_\omega = \dim\, \mathcal{V} = n(k+1)$.

Note that for $k = 1$ the above definition reduces to that of a symplectic vector space with a given Lagrangian subspace $\mathcal{W}$.[1]

Example 2.1. We consider the vector space $\mathcal{V} = \mathbb{R}^3$ with the family of skew-symmetric bilinear forms

$$\omega^1 = e^1 \wedge e^3 \quad \text{and} \quad \omega^2 = e^2 \wedge e^3\,,$$

and the subspace

$$\mathcal{W} = span\{e_1, e_2\},$$

where $\{e_1, e_2, e_3\}$ is the canonical basis of $\mathbb{R}^3$ and $\{e^1, e^2, e^3\}$ its dual basis. It is easy to check that

$$\omega^\alpha\big|_{\mathcal{W}\times\mathcal{W}} = 0\,, \quad \alpha = 1, 2\,.$$

[1] A subspace $\mathcal{W}$ of $\mathcal{V}$ is called a Lagrangian subspace if $\mathcal{W} \subset \mathcal{W}^\perp$, there exists another subspace $\mathcal{U}$ such that $\mathcal{U} \subset \mathcal{U}^\perp$ and $\mathcal{V} = \mathcal{W} \oplus \mathcal{U}$, (for more details see [de León and Vilariño (2012)]).

Moreover,

$$\ker\,\omega^1 = span\{e_2\} \text{ and } \ker\,\omega^2 = span\{e_1\}$$

and therefore $\ker\,\omega^1 \cap \ker\,\omega^2 = \{0\}$, that is, $(\omega^1, \omega^2, \mathcal{W})$ is a 2-symplectic structure on $\mathbb{R}^3$.

Example 2.2. We consider the vector space $\mathcal{V} = \mathbb{R}^6$ with the subspace

$$\mathcal{W} = span\{e_1, e_2, e_4, e_5\}$$

and the family of skew-symmetric bilinear forms

$$\omega^1 = e^1 \wedge e^3 + e^4 \wedge e^6 \text{ and } \omega^2 = e^2 \wedge e^3 + e^5 \wedge e^6$$

where $\{e_1, e_2, e_3, e_4, e_5, e_6\}$ is the canonical basis of $\mathbb{R}^6$ and the set $\{e^1, e^2, e^3, e^4, e^5, e^6\}$ the dual basis. It is easy to check that

$$\ker\,\omega^1 = span\{e_2, e_5\} \text{ and } \ker\,\omega^2 = span\{e_1, e_4\}$$

and therefore $\ker\,\omega_1 \cap \ker\,\omega_2 = \{0\}$. Moreover

$$\omega^\alpha\big|_{\mathcal{W}\times\mathcal{W}} = 0\,, \quad \alpha = 1, 2\,.$$

That is, $(\omega^1, \omega^2, \mathcal{W})$ is a 2-symplectic structure on $\mathbb{R}^6$.

Another k-symplectic structure on $\mathbb{R}^6$ is given by the family of 2-forms $\omega^\alpha = e^\alpha \wedge e^6$, with $1 \leq \alpha \leq 5$, and $\mathcal{W} = span\{e_1, e_2, e_3, e_4, e_5\}$ which is a 5-symplectic structure on $\mathbb{R}^6$.

Example 2.3. It is well known that for any vector space V, the space $V \times V^*$ admits a canonical symplectic form ω_V given by

$$\omega_V\left((v, \nu), (w, \eta)\right) = \eta(v) - \nu(w)\,,$$

for $v, w \in V$ and $\nu, \eta \in V^*$ (see for instance [Abraham and Marsden (1978)]). This structure has the following natural extension to the k-symplectic setting. For any k, the space $\mathcal{V} = V \times V^* \times \overset{k}{\cdots} \times V^*$ can be equipped with a family of k canonical skew-symmetric bilinear forms $(\omega_V^1, \dots, \omega_V^k)$ given by

$$\omega_V^\alpha\left((v, \nu_1, \dots, \nu_k), (w, \eta_1, \dots, \eta_k)\right) = \eta_\alpha(v) - \nu_\alpha(w)\,, \tag{2.12}$$

for $v, w \in V$ and $(\nu_1, \dots, \nu_k), (\eta_1, \dots, \eta_k) \in V^* \times \overset{k}{\cdots} \times V^*$. Now if we consider the subspace $\mathcal{W} = \{0\} \times V^* \times \overset{k}{\dots} \times V^*$ a simple computation shows that $(V \times V^* \times \overset{k}{\dots} \times V^*, \omega_V^1, \dots, \omega_V^k, \mathcal{W})$ is a k-symplectic vector space. In fact, this is a direct consequence of the computation of the kernel of ω_V^α for $1 \leq \alpha \leq k$, i.e.,

$$\ker \omega_V^\alpha = \{(v, \nu_1, \dots, \nu_k) \in \mathcal{V} \,|\, v = 0 \text{ and } \nu_\alpha = 0\}\,.$$

Let us observe that if we consider the natural projection

$$\begin{aligned} pr_\alpha \colon \; V \times V^* \times \overset{k}{\dots} \times V^* &\to V \times V^* \\ (v, \nu_1, \dots, \nu_k) \quad &\mapsto (v, \nu_\alpha), \end{aligned}$$

then the 2-form ω_V^α is exactly $(pr_\alpha)^* \omega_V$.

Definition 2.3. Let $(\mathcal{V}_1, \omega_1^1, \dots, \omega_1^k, \mathcal{W}_1)$ and $(\mathcal{V}_2, \omega_2^1, \dots, \omega_2^k, \mathcal{W}_2)$ be two k-symplectic vector space and let $\phi \colon \mathcal{V}_1 \to \mathcal{V}_2$ be a linear isomorphism. The map ϕ is called a ***k-symplectomorphism*** if it preserves the k-symplectic structure, that is

(1) $\phi^* \omega_2^\alpha = \omega_1^\alpha$; for each $1 \leq \alpha \leq k$,
(2) $\phi(\mathcal{W}_1) = \mathcal{W}_2$.

An important property of the k-symplectic structures is the following proposition, which establish a theorem of type Darboux for this generalization of the symplectic structure. A proof of the following result can be found in [Awane (1992); de León and Vilariño (2012)].

Prop 2.1. Let $(\omega^1, \dots, \omega^k, \mathcal{W})$ be a k-symplectic structure on the vector space $\mathcal{V}$. Then there exists a basis (Darboux basis) (e^i, f_i^α) of $\mathcal{V}$ (with $1 \leq i \leq n$ and $1 \leq \alpha \leq k$), such that for each $1 \leq \alpha \leq k$

$$\omega^\alpha = e^i \wedge f_i^\alpha \,.$$

2.2.2 *k-symplectic manifolds*

We turn now to the globalization of the ideas of the previous section to k-symplectic manifolds.

Definition 2.4. Let M be a smooth manifold of dimension $n(k+1)$, V be an integrable distribution of dimension nk and $\omega^1, \dots, \omega^k$ a family of closed differentiable 2-forms defined on M. In such a case $(\omega^1, \dots, \omega^k, V)$ is called ***a k-symplectic structure on*** M if and only if

(1) $\omega^\alpha \Big|_{V \times V} = 0, \quad 1 \leq \alpha \leq k\,,$

(2) $\displaystyle\bigcap_{\alpha=1}^{k} \ker \omega^\alpha = \{0\}\,.$

A manifold M endowed with a k-symplectic structure is said to be a ***k-symplectic manifold.***

Remark 2.3. In the above definition, the condition $\dim M = n(k+1)$ with $n, k \in \mathbb{N}$ implies that, for an arbitrary manifold M of dimension m, only a k-symplectic structure can exist if there is a couple (n,k) such that $M = n(k+1)$. Thus, for instance, if $M = \mathbb{R}^6$ there is no 3-symplectic structure for instance; in fact, only k-symplectic structures can exist if $k \in \{1, 2, 5\}$. ◇

Definition 2.5. Let $(M_1, \omega_1^1, \ldots, \omega_1^k, V_1)$ and $(M_2, \omega_2^1, \ldots, \omega_2^k, V_2)$ be two k-symplectic manifolds and let $\phi \colon M_1 \to M_2$ be a diffeomorphism. ϕ is called a ***k*-symplectomorphism** if it preserves the k-symplectic structure, that is if

(1) $\phi^* \omega_2^\alpha = \omega_1^\alpha$; for each $1 \leq \alpha \leq k$,
(2) $\phi_*(V_1) = V_2$.

Remark 2.4. Note that if $(M, \omega^1, \ldots, \omega^k, V)$ is a k-symplectic manifold then $(T_x M, \omega^1(x), \ldots, \omega^k(x), T_x V)$ is a k-symplectic vector space for all $x \in M$. ◇

Example 2.4. Let $(T_k^1)^* Q$ be the cotangent bundle of k^1-covelocities, then from (2.8) and (2.9) one easy checks that $(T_k^1)^* Q$, equipped with the canonical forms and the distribution $V = \ker(\pi^k)_*$, is a k-symplectic manifold.

Remark 2.5. For each $\nu_q \in (T_k^1)^* Q = T^*Q \oplus \stackrel{k}{\cdots} \oplus T^*Q$, the k-symplectic vector space $(T_{\nu_q}((T_k^1)^*Q),\ \omega^1(\nu_q), \ldots, \omega^k(\nu_q), T_{\nu_q} V)$ associated to the k-symplectic manifold $((T_k^1)^*Q, \omega^1, \ldots, \omega^k, V)$ is k-symplectomorphic to the canonical k-symplectic structure on $T_q Q \times T_q^* Q \times \stackrel{k}{\cdots} \times T_q^* Q$ described in example 2.3 with $V = T_q Q$. ◇

The following theorem is the differentiable version of Theorem 2.1. This theorem has been proved in [Awane (1992); de León, Méndez and Salgado (1988b)].

Theorem 2.1 (**k-symplectic Darboux theorem**). *Let $(M, \omega^1, \ldots, \omega^k, V)$ be a k-symplectic manifold. For every point of M we can find a local coordinate system (x^i, y_i^α), $1 \leq i \leq n$, $1 \leq \alpha \leq k$, called adapted coordinate system, such that*

$$\omega^\alpha = \sum_{i=1}^n dx^i \wedge dy_i^\alpha$$

for each $1 \leq \alpha \leq k$, *and*

$$V = span\Big\{\frac{\partial}{\partial y_i^\alpha}, 1 \leq i \leq n, 1 \leq \alpha \leq k\Big\}.$$

Remark 2.6. Notice that the notion of k-symplectic manifold introduced in this chapter coincides with the one given by A. Awane [Awane (1992); Awane and Goze (2000)], and it is equivalent to the notion of standard polysymplectic structure[2] of C. Günter [Günther (1987)] and integrable p-almost cotangent structure introduced by M. de León *et al.* [de León, Méndez and Salgado (1988,b)].

Observe that when $k = 1$, Awane's definition reduces to the notion of polarized symplectic manifold, that is a symplectic manifold with a Lagrangian submanifold. For that, in [de León and Vilariño (2012)] we distinguish between *k-symplectic* and *polarized k-symplectic manifolds.*

By taking a basis $\{e^1, \ldots, e^k\}$ of $\mathbb{R}^k$, every k-symplectic manifold $(N, \omega^1, \ldots, \omega^k)$ gives rise to a polysymplectic manifold $(N, \Omega = \sum_{i=1}^k \omega^i \otimes e_i)$. As Ω depends on the chosen basis, the polysymplectic manifold (N, Ω) is not canonically constructed. Nevertheless, two polysymplectic forms Ω_1 and Ω_2 induced by the same k-symplectic manifold and different bases for $\mathbb{R}^k$ are the same up to a change of basis on $\mathbb{R}^k$. In this case, we say that Ω_1 and Ω_2 are *gauge equivalent.* In a similar way, we say that $(N, \omega^1, \ldots, \omega^k)$ and $(N, \tilde{\omega}^1, \ldots, \tilde{\omega}^k)$ are gauge equivalent if they give rise to gauge equivalent polysymplectic forms, [Lucas and Vilariño (2015)].

$\diamond$

[2] A k-polysymplectic form on an $n(k+1)$-dimensional manifold N is an $\mathbb{R}^k$-valued closed nondegenerated 2-form on N of the form

$$\Omega = \sum_{i=1}^k \eta^i \otimes e_i\,,$$

where $\{e_1, \ldots, e_k\}$ is any basis of $\mathbb{R}^k$. The pair (N, Ω) is called a k-polysymplectic manifold.

Chapter 3

k-symplectic Formalism

In this chapter we shall describe the k-symplectic formalism. As we shall see in the following chapters, using this formalism we can study classical field theories in the Hamiltonian and Lagrangian cases.

One of the most important elements in the k-symplectic approach is the notion of k-vector field. Roughly speaking, it is a family of k vector fields. In order to introduce this notion in section 3.1, we previously consider the tangent bundle of k^1-velocities of a manifold, i.e., the Whitney sum of k copies of its tangent bundle with itself. In section 6.1 we shall describe this manifold with more details.

Here we shall introduce a geometric equation, called the k-symplectic Hamiltonian equation, which allows us to describe classical field theories when the k-symplectic manifold is the cotangent bundle of k^1-covelocities or its Lagrangian counterpart under some regularity condition satisfied by the Lagrangian function.

3.1 k-vector fields and integral sections

We shall devote this section to introduce the notion of k-vector field and discuss its integrability. This notion is fundamental in the k-symplectic and k-cosymplectic approaches.

Consider the tangent bundle $\tau\colon TM \to M$ of an arbitrary n-dimensional smooth manifold M and consider the space.[1]

$$T^1_kM = TM \oplus \overset{k}{\ldots} \oplus TM \,,$$

as the Whitney sum of k copies of the tangent bundle TM. Let us observe that an element v_p of T^1_kM is a family of k tangent vectors $(v_{1p}, \dots, v_{kp})$

[1] A complete description of this space T^1_kM can be found in section 6.1.

at the same point $p \in M$. Thus one can consider the canonical projection

$$\begin{aligned} \tau^k \colon \quad & T^1_k M \quad \to M \\ & \mathrm{v}_p = (v_{1p}, \dots, v_{kp}) \mapsto p\,. \end{aligned} \tag{3.1}$$

Definition 3.1. A ***k*-vector field** $\mathbf{X}$ on M is a section of the canonical projection $\tau^k \colon T^1_k M \to M$. We denote by $\mathfrak{X}^k(M)$ the set of k-vector fields on M.

Since $T^1_k M$ is the Whitney sum $TM \oplus \overset{k}{\dots} \oplus TM$ of k copies of TM, a k-vector field $\mathbf{X}$ on M defines a family of k vector fields $(X_1, \dots, X_k)$ on M through the projection of $\mathbf{X}$ onto every factor TM of the $T^1_k M$, as it is showed in the following diagram for each $1 \le \alpha \le k$:

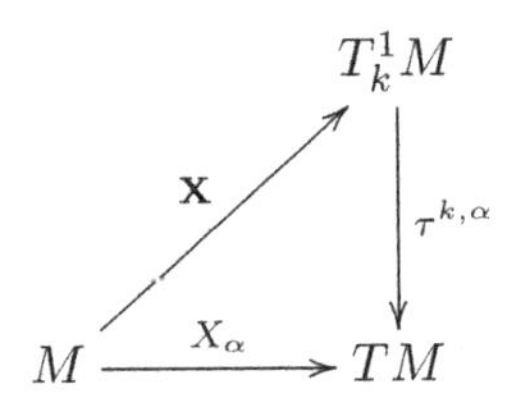

where $\tau^{k,\alpha}$ denotes the canonical projection over the αth component of $T^1_k M$, i.e.,

$$\begin{aligned} \tau^{k,\alpha} \colon \quad & T^1_k M \quad \to TM \\ & (v_{1p}, \dots, v_{kp}) \mapsto v_{\alpha p} \end{aligned} \,.$$

In what follows, we shall use indistinctly the notation $\mathbf{X}$ or $(X_1, \dots, X_k)$ to refer to a k-vector field.

Let us recall that given a vector field, we can consider the notion of integral curve. In this new setting we now introduce the generalization of this concept for k-vector fields: integral sections of a k-vector field.

Definition 3.2. An ***integral section*** of a k-vector field $\mathbb{X} = (X_1, \dots, X_k)$, passing through a point $p \in M$, is a map $\varphi : U_0 \subset \mathbb{R}^k \to M$, defined in some neighborhood U_0 of $0 \in \mathbb{R}^k$ such that

$$\varphi(0) = p, \ \varphi_*(x)\Big(\frac{\partial}{\partial x^\alpha}\Big|_x\Big) = X_\alpha(\varphi(x))\,, \tag{3.2}$$

for all $x \in U_0$ and for all $1 \le \alpha \le k$.

If there exists an integral section passing through each point of M, then $(X_1, \dots, X_k)$ is called an ***integrable k-vector field***.

Using local coordinates (U, y^i) on M we can write

$$\varphi_*(x)\left(\frac{\partial}{\partial x^\alpha}\Big|_x\right) = \frac{\partial \varphi^i}{\partial x^\alpha}\Big|_x \frac{\partial}{\partial y^i}\Big|_{\varphi(x)}, \quad X_\alpha = X_\alpha^i \frac{\partial}{\partial y^i}$$

where $\varphi^i = y^i \circ \varphi$.

Thus φ is an integral section of $\mathbf{X} = (X_1, \ldots, X_k)$ if and only if the following system of partial differential equations holds:

$$\frac{\partial \varphi^i}{\partial x^\alpha}\Big|_x = X_\alpha^i(\varphi(x)) \tag{3.3}$$

where $x \in U_0 \subseteq \mathbb{R}^k$, $1 \leq \alpha \leq k$ and $1 \leq i \leq n$.

Let us remark that if φ is an integral section of a k-vector field $\mathbf{X} = (X_1, \ldots, X_k)$, then each curve on M defined by $\varphi_\alpha(s) = \varphi(se_\alpha)$, with $\{e_1, \ldots, e_k\}$ the canonical basis on $\mathbb{R}^k$ and $s \in \mathbb{R}$, is an integral curve of the vector field X_α on M. However, given k integral curves of $X_1, \ldots, X_k$ respectively, it is not possible in general to reconstruct an integral section of $(X_1, \ldots, X_k)$.

We remark that a k-vector field $\mathbf{X} = (X_1, \ldots, X_k)$ with $\{X_1, \ldots, X_k\}$ linearly independent, is integrable if and only if $[X_\alpha, X_\beta] = 0$, for each α, β, that is, $\mathbf{X}$ is integrable if and only if the distribution generated by $\{X_1, \ldots, X_k\}$ is integrable. This is the geometric expression of the integrability condition of the preceding differential equation (see, for instance, [Dieudonné (1969)] or [Lee (2003)]).

Remark 3.1. k-vector fields in a manifold M can also be defined in a more general way as sections of the bundle $\Lambda^k M \to M$ (i.e., the contravariant skew-symmetric tensors of order k in M). Starting from the k-vector fields $\mathbf{X} = (X_1, \ldots, X_k)$ defined in Definition 3.1, and making the wedge product $X_1 \wedge \ldots \wedge X_k$, we obtain the particular class of the so-called *decomposable* or *homogeneous k-vector fields*, which can be associated with distributions on M. (See [Echeverría-Enríquez, Muñoz-Lecanda and Román-Roy (1998)] for a detailed exposition on these topics.) ◇

Example 3.1. Consider $M = (T_3^1)^*\mathbb{R}$ and a 3-vector field (X_1, X_2, X_3) with local expression

$$X_\alpha = p^\alpha \frac{\partial}{\partial q} + (X_\alpha)^\beta \frac{\partial}{\partial p^\beta}, \quad 1 \leq \alpha \leq 3,$$

where the functions $(X_\alpha)^\beta$ with $1 \leq \alpha, \beta \leq 3$ satisfy

$$(X_1)^1 + (X_2)^2 + (X_3)^3 = -4\pi r$$

r being a constant.

Then $\varphi\colon U_0 \to (T^1_3)^*\mathbb{R}$ with components $\varphi(x) = (\psi(x), \psi^\alpha(x))$ is an integral section of (X_1, X_2, X_3) if and only if (see (3.3))

$$\psi^\alpha = \frac{\partial \psi}{\partial x^\alpha}, \quad \alpha = 1, 2, 3,$$

$$4\pi r = -\Big(\frac{\partial \psi^1}{\partial x^1} + \frac{\partial \psi^2}{\partial x^2} + \frac{\partial \psi^3}{\partial x^3}\Big),$$

which are the electrostatic equations (for more details about these equations, see section 7.1).

3.2 k-symplectic Hamiltonian equation

Let $(M, \omega^1, \dots, \omega^k, V)$ a k-symplectic manifold and H a Hamiltonian function defined on M, that is, a function $H : M \to \mathbb{R}$.

Definition 3.3. The family (M, ω^α, H) is called ***k-symplectic Hamiltonian system***.

Given a k-symplectic Hamiltonian system (M, ω^α, H), we define a vector bundle morphisms $\flat_\omega$ as follows:

$$\begin{aligned} \flat_\omega\colon \quad & T^1_k M \quad \to T^*M \\ & (v_1, \dots, v_k) \mapsto \flat_\omega(v_1, \dots, v_k) = \mathrm{trace}(\iota_{v_\beta}\omega^\alpha) = \sum_{\alpha=1}^{k} \iota_{v_\alpha}\omega^\alpha . \end{aligned} \tag{3.4}$$

The above morphism induces a morphism of $\mathcal{C}^\infty(M)$-modules between the corresponding space of sections $\flat_\omega\colon \mathfrak{X}^k(M) \to \bigwedge^1(M)$.

Lemma 3.1. *The map $\flat_\omega$ is surjective.*

Proof. This result is a particular case of the following algebraic assertion: *If V is a vector space with a k-symplectic structure $(\omega^1, \dots, \omega^k, \mathcal{W})$, then the map*

$$\begin{aligned} \flat_\omega\colon \ & V \times \overset{k}{\dots} \times V \to V^* \\ & (v_1, \dots, v_k) \mapsto \flat_\omega(v_1, \dots, v_k) = \mathrm{trace}(\iota_{v_\beta}\omega^\alpha) = \sum_{\alpha=1}^{k} \iota_{v_\alpha}\omega^\alpha \end{aligned}$$

is surjective.

Indeed, we consider the identification

$$\begin{aligned} F\colon\ V^*\times \overset{k}{\ldots} \times V^* &\cong (V\times \overset{k}{\ldots} \times V)^* \\ (\nu_1,\ldots,\nu_k) &\mapsto F(\nu_1,\ldots,\nu_k)\,, \end{aligned} \tag{3.5}$$

where $F(\nu_1,\ldots,\nu_k)(v_1,\ldots,v_k) = \text{trace}\,\big(\nu_\alpha(v_\beta)\big) = \sum_{\alpha=1}^{k} \nu_\alpha(v_\alpha)$, and we consider the map $\sharp_\omega$ defined in (2.11).

We recall that as $(\omega^1,\ldots,\omega^k,\mathcal{W})$ is a k-symplectic structure, $\sharp_\omega$ is injective and therefore the dual map $\sharp_\omega^*$ is surjective.

Finally, using the identification (3.5) it is immediate to prove that $\flat_\omega = -\sharp_\omega^*$ and thus $\flat_\omega$ is surjective. □ □

Let (M,ω^α,H) be a k-symplectic Hamiltonian system and $\mathbf{X} \in \mathfrak{X}^k(M)$ a k-vector field solution of the geometric equation

$$\flat_\omega(\mathbf{X}) = \sum_{\alpha=1}^{k} \iota_{X_\alpha}\omega^\alpha = dH\,. \tag{3.6}$$

Given a local coordinate system $\big(q^i,p^\alpha_i\big)$, each X_α is locally given by

$$X_\alpha \;=\; (X_\alpha)^i\,\frac{\partial}{\partial q^i} \;+\; (X_\alpha)^\beta_i\,\frac{\partial}{\partial p^\beta_i}\,,\qquad 1\leq \alpha\leq k\,.$$

Now, since

$$dH \;=\; \frac{\partial H}{\partial q^i}\,dq^i \;+\; \frac{\partial H}{\partial p^\alpha_i}\,dp^\alpha_i$$

and

$$\omega^\alpha = dq^i \wedge dp^\alpha_i$$

we deduce that equation (3.6) is locally equivalent to the following equations

$$\frac{\partial H}{\partial q^i} \;=\; -\sum_{\beta=1}^{k}(X_\beta)^\beta_i\,,\qquad \frac{\partial H}{\partial p^\alpha_i} \;=\; (X_\alpha)^i\,, \tag{3.7}$$

with $1\leq i\leq n$ and $1\leq \alpha\leq k$.

Let us suppose now that the k-vector field $\mathbf{X} = (X_1,\ldots,X_k)$, solution of (3.6), is integrable and

$$\begin{aligned} \varphi : \mathbb{R}^k &\longrightarrow M \\ x &\rightarrow \varphi(x) = \big(\psi^i(x),\psi^\alpha_i(x)\big) \end{aligned}$$

is an integral section of $\mathbf{X}$, i.e., φ satisfies (3.2) which in this case is locally equivalent to the following system of partial differential equations (condition (3.3))

$$\left.\frac{\partial \psi^i}{\partial x^\alpha}\right|_x = (X_\alpha)^i(\varphi(x)), \quad \left.\frac{\partial \psi_i^\beta}{\partial x^\alpha}\right|_x = (X_\alpha)_i^\beta(\varphi(x)). \tag{3.8}$$

From (3.7) and (3.8) we obtain

$$\left.\frac{\partial H}{\partial q^i}\right|_{\varphi(x)} = -\sum_{\beta=1}^{k} \left.\frac{\partial \psi_i^\beta}{\partial x^\beta}\right|_x, \quad \left.\frac{\partial H}{\partial p_i^\alpha}\right|_{\varphi(x)} = \left.\frac{\partial \psi^i}{\partial x^\alpha}\right|_x \tag{3.9}$$

where $1 \leq i \leq n$, $1 \leq \alpha \leq k$.

This theory can be summarized as follows

Theorem 3.1. *Let (M, ω^α, H) be a k-symplectic Hamiltonian system and $\mathbf{X} = (X_1, \ldots, X_k)$ an integrable k-vector field on M solution of the equation (3.6).*

If $\varphi : \mathbb{R}^k \to M$ is an integral section of $\mathbf{X}$, then φ is a solution of the following systems of partial differential equations

$$\left.\frac{\partial H}{\partial q^i}\right|_{\varphi(x)} = -\sum_{\beta=1}^{k} \left.\frac{\partial \psi_i^\beta}{\partial x^\beta}\right|_x, \quad \left.\frac{\partial H}{\partial p_i^\alpha}\right|_{\varphi(x)} = \left.\frac{\partial \psi^i}{\partial x^\alpha}\right|_x.$$

From now, we shall call this equation (3.6) as ***k-symplectic Hamiltonian equation***.

Definition 3.4. A k-vector field $\mathbf{X} = (X_1, \ldots, X_k) \in \mathfrak{X}^k(M)$ is called a ***k-symplectic Hamiltonian k-vector field*** for a k-symplectic Hamiltonian system (M, ω^α, H) if $\mathbf{X}$ is a solution of (3.6). We denote by $\mathfrak{X}^k_H(M)$ the set of k-vector fields which are solution of (3.6), i.e.,

$$\mathfrak{X}^k_H(M) := \{\mathbf{X} = (X_1, \ldots, X_k) \in \mathfrak{X}^k(M) \,|\, \flat_\omega(\mathbf{X}) = dH\}. \tag{3.10}$$

One can guarantee the existence of the solution of the k-symplectic Hamiltonian equation (3.6), but the solution is not unique. In fact, let $H \in \mathcal{C}^\infty(M)$ be a function on M. As $dH \in \Omega^1(M)$ and the map $\flat_\omega$ is surjective, then there exists a k-vector field $\mathbf{X}^H = (X_1^H, \ldots, X_k^H)$ satisfying

$$\flat_\omega(X_1^H, \ldots, X_k^H) = dH, \tag{3.11}$$

i.e., $(X_1^H, \ldots, X_k^H)$ is a k-vector field solution of the k-symplectic Hamiltonian equation (3.6).

For instance one can define $\mathbf{X} = (X_1, \dots, X_k)$ locally as

$$\begin{aligned} X_1 &= \frac{\partial H}{\partial p_i^1}\frac{\partial}{\partial q^i} - \frac{\partial H}{\partial q^i}\frac{\partial}{\partial p_i^1} \\ X_\alpha &= \frac{\partial H}{\partial p_i^\alpha}\frac{\partial}{\partial q^i}, \quad 2 \leq \alpha \leq k \end{aligned} \tag{3.12}$$

and using a partition of the unity one can find a k-vector field $\mathbf{X} = (X_1, \dots, X_k)$ defined globally and satisfying (3.6).

Now we can assure the existence of solutions of (3.6) but not its uniqueness. In fact, let us observe that given a particular solution $(X_1, \dots, X_k)$ then any element of the set $(X_1, \dots, X_k) + \ker \flat_\omega$ is also a solution, since given $(Y_1, \dots, Y_k) \in \ker \flat_\omega$ then we have

$$Y_\beta^i = 0, \quad \sum_{\alpha=1}^{k} (Y_\alpha)_i^\alpha = 0\,, \tag{3.13}$$

where each Y_α is locally given by

$$Y_\alpha = Y_\alpha^i \frac{\partial}{\partial q^i} + (Y_\alpha)_i^\beta \frac{\partial}{\partial p_i^\beta}\,,$$

for $1 \leq \alpha \leq k$.

Another interesting remark is that a k-vector field solution of equation (3.6) is not necessarily integrable but in order to obtain the result of Theorem 3.1 the existence of integral sections is necessary. We recall that an integrable k-vector field is equivalent to the condition $[X_\alpha, X_\beta] = 0$ for all $1 \leq \alpha, \beta \leq k$.

Remark 3.2. Using the k-symplectic formalism presented in this chapter we can study symmetries and conservation laws on first-order classical field theories, see [Román-Roy, Salgado and Vilariño (2007, 2011)]. A large part of the discussion of the paper [Román-Roy, Salgado and Vilariño (2007)] is a generalization of the results obtained for non-autonomous mechanical systems (see, in particular, [Marmo, Saletan, Simoni and Vitale (1985); de León and Martín de Diego (1996b)]). The general problem of a group of symmetries acting on a k-symplectic manifold and the subsequent theory of reduction has been analyzed in [Marrero, Román-Roy, Salgado and Vilariño (2014); Munteanu, Rey and Salgado (2004)]. We further remark that the problem of symmetries in field theory has been analyzed using other geometric frameworks, see for instance [Echeverría-Enríquez, Muñoz-Lecanda and Román-Roy (1999); Gotay, Isenberg, Marsden and Montgomery (2004);

Gotay, Isenberg, Marsden (2004); de León, Martín de Diego and Santamaría-Merino (2004)]. About this topic, Noether's theorem associates conservation laws to Cartan symmetries, however, these kinds of symmetries do not exhaust the set of symmetries. Different attempts have been made to extend Noether's theorem in order to include the so-called hidden symmetries and the corresponding conserved quantities, see for instance [Sarlet and Cantrijn (1981)] in mechanics, [Echeverría-Enríquez, Muñoz-Lecanda and Román-Roy (1999)] in multisymplectic field theories or [Román-Roy, Salgado and Vilariño (2013)] in the k-symplectic setting.

The k-symplectic formalism described here can be extended to other geometrical approaches. For instance:

- The k-symplectic approach can also be studied when one considers classical field theories subject to nonholonomic constraints [de León, Martín de Diego, Salgado and Vilariño (2008)]. The procedure developed in [de León, Martín de Diego, Salgado and Vilariño (2008)] extends that by Bates and Sniatycki [Bates and Sniatycki (1993)] for the linear case. The interest of the study of nonholonomic constraints has been stimulated by its close connection to problems in control theory (see, for instance, [Bloch (2003); Bloch, Krishnaprasad, Marsden and Murray (1996); Cortés (2002)]. In the literature, one can distinguish mainly two different approaches in the study of systems subjected to a nonholonomic constraints. The first one is based on the d'Alembert's principle and the second is a constrained variational approach. As is well known, the dynamical equation generated by both approaches are in general not equivalent [Cortés, de León, Martín de Diego and Martínez (2003)]. The nonholonomic field theory has been studied using another geometrical approaches, (see, for instance [Binz, de León, Martín de Diego and Socolescu (2002); de León, Marrero and Martín de Diego (1997,b); de León and Martín de Diego (1996); de León, Martín de Diego and Santamaría-Merino (2004b); Vankerschaver (2007,b); Vankerschaver, Cantrijn, de León and Martín de Diego (2005); Vankerschaver and Martín de Diego (2008)]).
- Another interesting setting is the category of the Lie algebroids [Mackenzie (1987, 1995)]. For further information on groupoids and Lie algebroids and their roles in differential geometry see [Cannas da Silva and Weinstein (1999); Higgins and Mackenzie (1990)]. Let us remember that a Lie algebroid is a generalization of both the Lie algebra and the integrable distribution. The idea of using Lie algebroids

in mechanics is due to Weinstein [Weinstein (1996)]. His formulation allows a geometric unified description of dynamical systems with a variety of different kinds of phase spaces: Lie groups, Lie algebras, Cartesian products of manifolds, quotients manifolds,.... Two good surveys of this topic are [Cortés, de León, Marrero, Martín de Diego and Martinez (2006); de León, Marrero and Martínez (2005)]. In [de León, Martín de Diego, Salgado and Vilariño (2009)] we describe the k-symplectic formalism on Lie algebroids.

- The Skinner-Rusk approach [Skinner and Rusk (1983)] can be considered in the k-symplectic formalism. This topic was studied in [Rey, Román-Roy and Salgado (2005)] in the k-symplectic approach and in [Rey, Román-Roy, Salgado and Vilariño (2012)] in the k-cosymplectic approach.
- Another interesting topic is the study of Lagrangian submanifolds in the k-symplectic setting [de León and Vilariño (2012)]. In this chapter, we extend the well-known normal form theorem for Lagrangian submanifolds proved by Weinstein in symplectic geometry to the setting of k-symplectic manifolds. ◇

3.3 Example: electrostatic equations

Consider the 3-symplectic Hamiltonian equations

$$\iota_{X_1}\omega^1 + \iota_{X_2}\omega^2 + \iota_{X_3}\omega^3 = dH\,, \tag{3.14}$$

where H is the Hamiltonian function given by

$$\begin{aligned} H: \quad (T^1_3)^*\mathbb{R} \quad &\longrightarrow \mathbb{R} \\ (q,p^1,p^2,p^3) \ \to \ & 4\pi r q + \frac{1}{2}\sum_{\alpha=1}^{3}(p^\alpha)^2\,. \end{aligned} \tag{3.15}$$

Let us observe that in this example the k-symplectic manifold is the cotangent bundle of 3-covelocities of the real line $(T^1_3)^*\mathbb{R}$ with its canonical 3-symplectic structure.

If (X_1, X_2, X_3) is a solution of (3.14) then, since

$$\frac{\partial H}{\partial q} = 4\pi r\,, \quad \frac{\partial H}{\partial p^\alpha} = p^\alpha\,,$$

and from (3.7) we deduce that each X_α, with $1 \leq \alpha \leq 3$, has the local expression

$$X_\alpha = p^\alpha\frac{\partial}{\partial q} + (X_\alpha)^\beta\frac{\partial}{\partial p^\beta},$$

where the function components $(X_\alpha)^\beta$ with $1 \leq \alpha, \beta \leq 3$ satisfy the identity

$$4\pi r = -\Big((X_1)^1 + (X_2)^2 + (X_3)^3 \Big) .$$

Let us suppose that (X_1, X_2, X_3) is integrable, that is, in this particular case, the functions $(X_\alpha)^\beta$ with $1 \leq \alpha, \beta \leq k$ satisfy

$$(X_\alpha)^\beta = (X_\beta)^\alpha$$

and

$$X_1((X_2)^\beta) = X_2((X_1)^\beta),\ X_1((X_3)^\beta) = X_3((X_1)^\beta),\ X_2((X_3)^\beta) = X_3((X_2)^\beta) .$$

Under the assumption of integrability of (X_1, X_2, X_3), if

$$\begin{aligned} \varphi : \mathbb{R}^3 &\longrightarrow (T^1_3)^*\mathbb{R} \\ x &\to \varphi(x) = (\psi(x), \psi^1(x), \psi^2(x), \psi^3(x)) \end{aligned}$$

is an integral section of a 3-vector field (X_1, X_2, X_3) solution of (3.14), then we deduce that

$$(\psi(x), \psi^1(x), \psi^2(x), \psi^3(x))$$

is a solution of

$$\begin{aligned} \psi^\alpha &= \frac{\partial \psi}{\partial x^\alpha}, \\ -\Big(\frac{\partial \psi^1}{\partial x^1} + \frac{\partial \psi^2}{\partial x^2} + \frac{\partial \psi^3}{\partial x^3} \Big) &= 4\pi r , \end{aligned} \tag{3.16}$$

which is a particular case of the electrostatic equations (for a more detail description of these equations, see section 7.1).

Chapter 4

Hamiltonian Classical Field Theory

In this chapter we shall study Hamiltonian classical field theories, that is, we shall discuss the Hamilton-De Donder-Weyl equations (these equations will also be called the HDW equations for short) which have the following local expression

$$\left.\frac{\partial H}{\partial q^i}\right|_{\varphi(x)} = -\sum_{\alpha=1}^{k} \left.\frac{\partial \psi_i^\alpha}{\partial x^\alpha}\right|_t, \quad \left.\frac{\partial H}{\partial p_i^\alpha}\right|_{\varphi(x)} = \left.\frac{\partial \psi^i}{\partial x^\alpha}\right|_x, \tag{4.1}$$

where $H\colon (T_k^1)^*Q \to \mathbb{R}$ is a Hamiltonian function.

A solution of these equations is a map

$$\begin{aligned} \varphi : \mathbb{R}^k &\longrightarrow (T_k^1)^*Q \\ x &\to \varphi(x) = (\psi^i(x), \psi_i^\alpha(x)) \end{aligned}$$

where $1 \leq i \leq n$, $1 \leq \alpha \leq k$.

In a classical view these equations can be obtained from a multiple integral variational problem. In this chapter we shall describe this variational approach and then we shall give a new geometric way of obtaining the HDW equations using the k-symplectic formalism described in section 2.2.2 when the k-symplectic manifolds is the canonical model of these structures: the manifold $((T_k^1)^*Q, \omega^1, \dots, \omega^k, V)$ described in section 2.1.

4.1 Variational approach

In Hamiltonian mechanics, the Hamilton equations are obtained from a variational principle. This can be generalized to classical field theory, where the problem consists of finding the extremal of a variational problem associated to multiple integrals of Hamiltonian densities.

In this subsection we shall see that the Hamilton-De Donder-Weyl equations (4.1) are equivalent a one variational principle on the space of smooth maps with compact support; we denote this set by $\mathcal{C}^\infty_C(\mathbb{R}^k, (T^1_k)^*Q)$.

To describe this variational principle we need the notion of prolongation of diffeomorphisms and vector fields from Q to the cotangent bundle of k^1-covelocities, which we shall introduce in the sequel.

4.1.1 *Prolongation of diffeomorphism and vector fields*

Given a diffeomorphism between two manifolds M and N we can consider an induced map between $(T^1_k)^*N$ and $(T^1_k)^*M$. This map allows us to define the prolongation of vector fields from Q to the cotangent bundle of k^1-covelocities.

Definition 4.1. Let $f\colon M \to N$ be a diffeomorphism. The ***natural or canonical prolongation*** of f to the corresponding bundles of k^1-covelocities is the map

$$(T^1_k)^*f : (T^1_k)^*N \to (T^1_k)^*M$$

defined as follows:

$$\begin{aligned}(T^1_k)^*f(\nu_{1\,f(x)},\dots,\nu_{k\,f(x)}) &= (f^*(\nu_{1\,f(x)}),\dots,f^*(\nu_{k\,f(x)}))\\ &= (\nu_{1\,f(x)}\circ f_*(x),\dots,\nu_{k\,f(x)}\circ f_*(x))\end{aligned}$$

where $(\nu_{1\,f(x)},\dots,\nu_{k\,f(x)}) \in (T^1_k)^*N$ and $m \in M$.

The canonical prolongation of diffeomorphism allows us to introduce the canonical or complete lift of vector fields from Q to $(T^1_k)^*Q$.

Definition 4.2. Let Z be a vector field on Q, with 1-parameter group of diffeomorphism $\{h_s\}$. The ***canonical or complete lift*** of Z to $(T^1_k)^*Q$ is the vector field Z^{C*} on $(T^1_k)^*Q$ whose local 1-parameter group of diffeomorphism is $\{(T^1_k)^*(h_s)\}$.

Let Z be a vector field on Q with local expression $Z = Z^i\dfrac{\partial}{\partial q^i}$. In the canonical coordinate system (2.2) on $(T^1_k)^*Q$, the local expression of Z^{C*} is

$$Z^{C*} = Z^i\frac{\partial}{\partial q^i} - p^\alpha_j\frac{\partial Z^j}{\partial q^k}\frac{\partial}{\partial p^\alpha_k}\,. \tag{4.2}$$

The canonical prolongation of diffeomorphisms and vector fields from Q to $(T^1_k)^*Q$ have the following properties.

Lemma 4.1.

(1) *Let $\varphi\colon Q \to Q$ be a diffeomorphism and $\Phi = (T^1_k)^*\varphi$ the canonical prolongation of φ to $(T^1_k)^*Q$. Then:*

$$(i)\ \Phi^*\theta^\alpha = \theta^\alpha \quad \textit{and} \quad (ii)\ \Phi^*\omega^\alpha = \omega^\alpha, \tag{4.3}$$

where $1 \leq \alpha \leq k$.

(2) *Let $Z \in \mathfrak{X}(Q)$ and Z^{C*} be the complete lift of Z to $(T^1_k)^*Q$. Then*

$$(i)\ \mathcal{L}_{Z^{C*}}\theta^\alpha = 0 \quad \textit{and} \quad (ii)\ \mathcal{L}_{Z^{C*}}\omega^\alpha = 0, \tag{4.4}$$

with $1 \leq \alpha \leq k$.

Proof. **(1)** (i) is a consequence of the commutativity of the following diagram

$$\begin{array}{ccc}
(T^1_k)^*Q & \xrightarrow{(T^1_k)^*\varphi} & (T^1_k)^*Q \\
\Big\downarrow \pi^{k,\alpha} & & \Big\downarrow \pi^{k,\alpha} \\
T^*Q & \xrightarrow{\varphi^*} & T^*Q
\end{array}$$

for each $1 \leq \alpha \leq k$, that is,

$$\pi^{k,\alpha} \circ (T^1_k)^*\varphi = \varphi^* \circ \pi^{k,\alpha}.$$

In fact, using the above identity one has

$$\begin{aligned}
\left[(T^1_k)^*\varphi\right]^* \theta^\alpha &= \left[(T^1_k)^*\varphi\right]^* ((\pi^{k,\alpha})^*\theta) = \left(\pi^{k,\alpha} \circ (T^1_k)^*\varphi\right)^*\theta \\
&= (\varphi^* \circ \pi^{k,\alpha})^*\theta = (\pi^{k,\alpha})^*((\varphi^*)^*\theta) = (\pi^{k,\alpha})^*\theta = \theta^\alpha ,
\end{aligned}$$

where we have used the identity $(\varphi^*)^*\theta = \theta$ (see [Abraham and Marsden (1978)], p. 180).

Item (ii) is a direct consequence of (i) and the definition of the closed 2-forms $\omega^1, \dots, \omega^k$.

(2) Since the infinitesimal generator of Z^{C*} is the canonical prolongation of the infinitesimal generator of Z, then from item (1) of this lemma one obtains that the second part holds.

□

4.1.2 *Variational principle*

Now we are in a position to describe the multiple integral problem from which one obtains the Hamilton-De Donder-Weyl equations.

We denote by d^kx the volume form on $\mathbb{R}^k$ given by $dx^1 \wedge \ldots \wedge dx^k$ and $d^{k-1}x_\alpha$ is the $(k-1)$-form defined by

$$d^{k-1}x_\alpha = \iota_{\partial/\partial x^\alpha} d^k x\,,$$

for each $1 \leq \alpha \leq k$.

Before describing the variational problem in this setting we recall the following result:

Lemma 4.2. *Let G denote a fixed simply-connected domain in the k-dimensional space, bounded by a hypersurface ∂G. If $\Phi(x)$ is a continuous function in G and if*

$$\int_G \Phi(x)\eta(x)d^kx = 0$$

for all function $\eta(x)$ of class C^1 which vanish on the boundary ∂G of G, then

$$\Phi(x) = 0$$

in G.

A proof of this lemma can be found in [Rund (1973)].

Definition 4.3. Denote by $C_C^\infty(\mathbb{R}^k, (T_k^1)^*Q)$ the set of maps

$$\varphi : U_0 \subseteq \mathbb{R}^k \to (T_k^1)^*Q,$$

with compact support defined on an open set U_0. Let $H : (T_k^1)^*Q \to \mathbb{R}$ be a Hamiltonian function, then we define the integral action associated to H by

$$\begin{aligned}\mathcal{H} : C_C^\infty(\mathbb{R}^k, (T_k^1)^*Q) &\to \mathbb{R}\\ \varphi \qquad &\mapsto \int_{\mathbb{R}^k} \Big(\sum_{\alpha=1}^{k} (\varphi^*\theta^\alpha) \wedge d^{k-1}x_\alpha - (\varphi^* H) d^k x \Big)\,.\end{aligned}$$

Definition 4.4. A map $\varphi \in C_C^\infty(\mathbb{R}^k, (T_k^1)^*Q)$ is an extremal of $\mathcal{H}$ if

$$\frac{d}{ds}\Big|_{s=0} \mathcal{H}(\tau_s \circ \varphi) = 0$$

for each flow τ_s on $(T_k^1)^*Q$ such that $\tau_s(\nu_{1_q}, \ldots, \nu_{k_q}) = (\nu_{1_q}, \ldots, \nu_{k_q})$ for all $(\nu_{1_q}, \ldots, \nu_{k_q})$ on the boundary of $\varphi(U_0) \subset (T_k^1)^*Q$, that is, we consider the variations of φ given by the composition by elements of one-parametric group of diffeomorphism which leaves invariant the boundary of the image of φ.

Let us observe that the flows $\tau_s : (T^1_k)^*Q \to (T^1_k)^*Q$ considered in the above definition are generated by vector fields on $(T^1_k)^*Q$ which are zero on the boundary of $\varphi(U_0)$.

The variational problem here considered consists of finding the extremals of the integral action $\mathcal{H}$.

The following proposition gives us a characterization of the extremals of the integral action $\mathcal{H}$ associated with the Hamiltonian H.

Prop 4.1. Let $H : (T^1_k)^*Q \to \mathbb{R}$ be a Hamiltonian function and $\varphi \in \mathcal{C}^\infty_C(\mathbb{R}^k, (T^1_k)^*Q)$. The following statements are equivalent:

(1) $\varphi : U_0 \subset \mathbb{R}^k \to (T^1_k)^*Q$ is an extremal of the variational problem associated to H.

(2) For each vector field Z on Q, such that its complete lift Z^{C*} to $(T^1_k)^*Q$ vanishes on the boundary of $\varphi(U_0)$, the equality

$$\int_{\mathbb{R}^k} \left([\varphi^*(\mathcal{L}_{Z^{C*}}\theta^\alpha)] \wedge d^{k-1}x_\alpha - [\varphi^*(\mathcal{L}_{Z^{C*}}H)]d^kx \right) = 0\,,$$

holds.

(3) φ is solution of the Hamilton-De Donder-Weyl equations, that is, if φ is locally given by $\varphi(x) = (\psi^i(x), \psi^\alpha_i(x))$, then the functions ψ^i, ψ^α_i satisfy the system of partial differential equations (4.1).

Proof. First we shall prove the equivalence between items **(1)** and **(2)** $(1 \Leftrightarrow 2)$.

Let $Z \in \mathfrak{X}(Q)$ be a vector field on Q satisfying the conditions in **(1)**, and with one-parameter group of diffeomorphism $\{\tau_s\}$. Then, from the definition of the complete lift we know that Z^{C*} generates the one-parameter group $\{(T^1_k)^*\tau_s\}$.

Thus,

$$\begin{aligned}
&\frac{d}{ds}\Big|_{s=0} \mathcal{H}((T^1_k)^*\tau_s \circ \varphi) \\
= \;& \frac{d}{ds}\Big|_{s=0} \int_{\mathbb{R}^k} \Big(\sum_{\alpha=1}^k ([(T^1_k)^*\tau_s \circ \varphi]^*\theta^\alpha) \wedge d^{k-1}x_\alpha - ([(T^1_k)^*\tau_s \circ \varphi]^*H)d^kx \Big) \\
= \;& \lim_{s\to 0} \frac{1}{s} \Big(\int_{\mathbb{R}^k} \Big(\sum_{\alpha=1}^k ([(T^1_k)^*\tau_s \circ \varphi]^*\theta^\alpha) \wedge d^{k-1}x_\alpha - ([(T^1_k)^*\tau_s \circ \varphi]^*H)d^kx \Big) \\
& - \int_{\mathbb{R}^k} \Big(\sum_{\alpha=1}^k ([(T^1_k)^*\tau_0 \circ \varphi]^*\theta^\alpha) \wedge d^{k-1}x_\alpha - ([(T^1_k)^*\tau_0 \circ \varphi]^*H)d^kx \Big) \Big)
\end{aligned}$$

$$
\begin{aligned}
= & \lim_{s\to 0}\frac{1}{s}\left(\int_{\mathbb{R}^k}\sum_{\alpha=1}^{k}([(T^1_k)^*\tau_s\circ\varphi]^*\theta^\alpha)\wedge d^{k-1}x_\alpha-\int_{\mathbb{R}^k}\sum_{\alpha=1}^{k}(\varphi^*\theta^\alpha)\wedge d^{k-1}x_\alpha\right)\\
& -\lim_{s\to 0}\frac{1}{s}\left(\int_{\mathbb{R}^k}([(T^1_k)^*\tau_s\circ\varphi]^*H)d^kx-\int_{\mathbb{R}^k}(\varphi^*H)d^kx\right)\\
= & \lim_{s\to 0}\frac{1}{s}\left(\int_{\mathbb{R}^k}\sum_{\alpha=1}^{k}[\varphi^*\Big(((T^1_k)^*\tau_s)^*\theta^\alpha-\theta^\alpha\Big)]\wedge d^{k-1}x_\alpha\right)\\
& -\lim_{s\to 0}\frac{1}{s}\left(\int_{\mathbb{R}^k}[\varphi^*\Big(((T^1_k)^*\tau_s)^*H-H\Big)]d^kx\right)\\
= & \int_{\mathbb{R}^k}([\varphi^*(\mathcal{L}_{Z^{C^*}}\theta^\alpha)]\wedge d^{k-1}x_\alpha-[\varphi^*(\mathcal{L}_{Z^{C^*}}H)]d^kx)\ ,
\end{aligned}
$$

where in the last identity we are using the definition of Lie derivative with respect to Z^{C^*}.

Therefore, φ is an extremal of $\mathcal{H}$ if and only if

$$\int_{\mathbb{R}^k}([\varphi^*(\mathcal{L}_{Z^{C^*}}\theta^\alpha)]\wedge d^{k-1}x_\alpha-[\varphi^*(\mathcal{L}_{Z^{C^*}}H)]d^kx)=0\,.$$

We now prove the equivalence between (**2**) and (**3**) ($2\Leftrightarrow 3$).

Taking into account that

$$\mathcal{L}_{Z^{C^*}}\theta^\alpha=d\iota_{Z^{C^*}}\theta^\alpha+\iota_{Z^{C^*}}d\theta^\alpha$$

one obtains

$$
\begin{aligned}
\int_{\mathbb{R}^k}[\varphi^*(\mathcal{L}_{Z^{C^*}}\theta^\alpha)]\wedge d^{k-1}x_\alpha = & \textstyle\int_{\mathbb{R}^k}[\varphi^*(d\iota_{Z^{C^*}}\theta^\alpha)]\wedge d^{k-1}x_\alpha\\
& +\textstyle\int_{\mathbb{R}^k}[\varphi^*(\iota_{Z^{C^*}}d\theta^\alpha)]\wedge d^{k-1}x_\alpha\,.
\end{aligned}
$$

Since

$$[\varphi^*(d\iota_{Z^{C^*}}\theta^\alpha)]\wedge d^{k-1}x_\alpha=d\Big(\varphi^*(\iota_{Z^{C^*}}\theta^\alpha)\wedge d^{k-1}x_\alpha\Big)$$

then $[\varphi^*(d\iota_{Z^{C^*}}\theta^\alpha)]\wedge d^{k-1}x_\alpha$ is a closed 1-form on $\mathbb{R}^k$. Therefore, applying Stoke's theorem one obtains:

$$\int_{\mathbb{R}^k}[\varphi^*(d\iota_{Z^{C^*}}\theta^\alpha)]\wedge d^{k-1}x_\alpha=\int_{\mathbb{R}^k}d\Big(\varphi^*(\iota_{Z^{C^*}}\theta^\alpha)\wedge d^{k-1}x_\alpha\Big)=0\,.$$

Then,

$$\int_{\mathbb{R}^k}([\varphi^*(\mathcal{L}_{Z^{C^*}}\theta^\alpha)]\wedge d^{k-1}x_\alpha-[\varphi^*(\mathcal{L}_{Z^{C^*}}H)]d^kx)=0$$

if and only if,

$$\int_{\mathbb{R}^k} \left([\varphi^* \left(\iota_{Z^{C*}} d\theta^\alpha \right)] \wedge d^{k-1}x_\alpha - [\varphi^*(\mathcal{L}_{Z^{C*}} H)] d^k x \right) = 0 \,.$$

Consider now the canonical coordinate system such that $Z = Z^i \dfrac{\partial}{\partial q^i}$; taking into account the local expression (4.2) for the complete lift Z^{C^*} and that $\varphi(x) = (\psi^i(x), \psi_i^\alpha(x))$, we have

$$\begin{aligned}
&\varphi^* \left(\iota_{Z^{C*}} d\theta^\alpha \right) \wedge d^{k-1}x_\alpha - \varphi^*(\mathcal{L}_{Z^{C*}} H) d^k x \\
&= -(Z^i(x)) \left(\sum_{\alpha=1}^k \frac{\partial \psi_i^\alpha}{\partial x^\alpha}\Big|_x + \frac{\partial H}{\partial q^i}\Big|_{\varphi(x)} \right) d^k x \\
&\quad - \left[\sum_{\alpha=1}^k \psi_j^\alpha(x) \frac{\partial Z^j}{\partial q^i}\Big|_x \left(\frac{\partial \psi^i}{\partial x^\alpha}\Big|_x - \frac{\partial H}{\partial p_i^\alpha}\Big|_{\varphi(x)} \right) \right] d^k x
\end{aligned}$$

for each $Z \in \mathfrak{X}(Q)$ (under the conditions established in this theorem), where we are using the notation $Z^i(x) := (Z^i \circ \pi^k \circ \varphi)(x)$. From the last expression we deduce that φ is an extremal of $\mathcal{H}$ if and only if

$$\begin{aligned}
&\int_{\mathbb{R}^k} Z^i(x) \left(\sum_{\alpha=1}^k \frac{\partial \psi_i^\alpha}{\partial x^\alpha}\Big|_x + \frac{\partial H}{\partial q^i}\Big|_{\varphi(x)} \right) d^k x \\
&+ \int_{\mathbb{R}^k} \sum_{\alpha=1}^k \psi_j^\alpha(x) \frac{\partial Z^j}{\partial q^i}\Big|_x \left(\frac{\partial \psi^i}{\partial x^\alpha}\Big|_x - \frac{\partial H}{\partial p_i^\alpha}\Big|_{\varphi(x)} \right) d^k x = 0
\end{aligned}$$

for all Z^i. Therefore,

$$\begin{aligned}
&\int_{\mathbb{R}^k} (Z^i(x)) \left(\sum_{\alpha=1}^k \frac{\partial \psi_i^\alpha}{\partial x^\alpha}\Big|_x + \frac{\partial H}{\partial q^i}\Big|_{\varphi(x)} \right) d^k x = 0 \\
&\int_{\mathbb{R}^k} \sum_{\alpha=1}^k \psi_j^\alpha(x) \frac{\partial Z^j}{\partial q^i}\Big|_x \left(\frac{\partial \psi^i}{\partial x^\alpha}\Big|_x - \frac{\partial H}{\partial p_i^\alpha}\Big|_{\varphi(x)} \right) d^k x = 0
\end{aligned} \tag{4.5}$$

for all $Z \in \mathfrak{X}(Q)$ satisfying the statements of this theorem, and, thus, for any values $Z^i(q)$ and $\dfrac{\partial Z^j}{\partial q^i}\Big|_q$.

Applying Lemma 4.2, from (4.5) one obtains that,

$$\sum_{\alpha=1}^k \frac{\partial \psi_i^\alpha}{\partial x^\alpha}\Big|_x + \frac{\partial H}{\partial q^i}\Big|_{\varphi(x)} = 0 \quad , \quad \sum_{\alpha=1}^k \psi_j^\alpha(x) \left(\frac{\partial \psi^i}{\partial x^\alpha}\Big|_x - \frac{\partial H}{\partial p_i^\alpha}\Big|_{\varphi(x)} \right) = 0 \,.$$

The first group of equations gives us the first group of the Hamilton-De Donder-Weyl equations (4.1).

Now, consider the second set of the above equations, it follows that

$$\frac{\partial \psi^i}{\partial x^\alpha}\Big|_x - \frac{\partial H}{\partial p_i^\alpha}\Big|_{\varphi(x)} = 0\,,$$

which is the second set of the Hamilton-De Donder-Weyl equations (4.1).

The converse is obtained starting from the Hamilton-De Donder-Weyl equations and reversing the arguments in the above proof. □

4.2 Hamilton-De Donder-Weyl equations

The above variational principle allows us to obtain the HDW equations but there exist other methods to obtain these equations: one of them consists of using the k-symplectic Hamiltonian equation when we consider the k-symplectic manifold $M = (T^1_k)^*Q$.

In this case, we take a Hamiltonian function $H \in \mathcal{C}^\infty((T^1_k)^*Q)$. Thus, from Theorem 3.1, one obtains that given an integrable k-vector field $\mathbf{X} = (X_1, \ldots, X_k) \in \mathfrak{X}^k_H((T^1_k)^*Q)$ and an integral section $\varphi\colon U \subset \mathbb{R}^k \to (T^1_k)^*Q$ of $\mathbf{X}$, φ is a solution of the following systems of partial differential equations

$$\frac{\partial H}{\partial q^i}\Big|_{\varphi(x)} = -\sum_{\beta=1}^{k} \frac{\partial \psi_i^\beta}{\partial x^\beta}\Big|_x\,, \quad \frac{\partial H}{\partial p_i^\alpha}\Big|_{\varphi(x)} = \frac{\partial \psi^i}{\partial x^\alpha}\Big|_x\,,$$

that is, φ is a solution of the HDW equations (4.1).

Therefore, given an integrable k-vector field $X \in \mathfrak{X}^k_H((T^1_k)^*Q)$, its integral sections are solutions of the HDW equations. Now it is natural to do the following question: *Given a solution $\varphi\colon U \subset \mathbb{R}^k \to (T^1_k)^*Q$ of the HDW equations, is there a k-vector field $X \in \mathfrak{X}^k_H((T^1_k)^*Q)$ such that φ is an integral section of $\mathbf{X}$?*

Here we give an answer to this question.

Prop 4.2. If a map $\varphi : \mathbb{R}^k \to (T^1_k)^*Q$ is a solution of the HDW equation (4.1) and φ is an integral section of an integrable k-vector field $\mathbf{X} \in \mathfrak{X}^k((T^1_k)^*Q)$, then $\mathbf{X} = (X_1, \ldots, X_k)$ is a solution of equation (3.6) at the points of the image of φ.

Proof. We must prove that

$$\frac{\partial H}{\partial p_i^\alpha}(\varphi(x)) = (X_\alpha)^i(\varphi(x)), \quad \frac{\partial H}{\partial q^i}(\varphi(x)) = -\sum_{\alpha=1}^{k}(X_\alpha)_i^\alpha(\varphi(x))\,. \qquad (4.6)$$

Now as $\varphi(x) = (\psi^i(x), \psi_i^\alpha(x))$ is an integral section of $\mathbf{X}$ we have that (3.8) holds; but, as φ is also a solution of the Hamilton-De Donder-Weyl equation (4.1), then we deduce (4.6). □

We cannot claim that $\mathbf{X} \in \mathfrak{X}^k_H((T^1_k)^*Q)$ because we cannot assure that $\mathbf{X}$ is a solution of equations (3.6) on the whole in $(T^1_k)^*Q$.

Remark 4.1. It is also important to point out that equations (4.1) and (3.6) are not equivalent in the sense that not every solution of (4.1) is an integral section of some integrable k-vector field belonging to $\mathfrak{X}^k_H((T^1_k)^*Q)$. ◇

Definition 4.5. A map $\varphi\colon \mathbb{R}^k \to (T^1_k)^*Q$, solution of equation (4.1), is said to be an ***admissible solution*** to the HDW-equation for a k-symplectic Hamiltonian system $((T^1_k)^*Q, \omega^\alpha, H)$ if it is an integral section of some integrable k-vector field $\mathbf{X} \in \mathfrak{X}^k((T^1_k)^*Q)$.

If we consider only admissible solutions to the HDW-equations of k-symplectic Hamiltonian systems, we say that $((T^1_k)^*Q, \omega^\alpha, H)$ is an ***admissible k-symplectic Hamiltonian system.***

In this way, for admissible k-symplectic Hamiltonian systems, the geometric field equation (3.6) for integrable k-vector fields is equivalent to the HDW-equation (4.1) (as it is established in Theorem 3.1 and Proposition 4.2).

Chapter 5

Hamilton-Jacobi Theory in k-symplectic Field Theories

The usefulness of Hamilton-Jacobi theory in classical mechanics is well-known, giving an alternative procedure to study and, in some cases, to solve the evolution equations [Abraham and Marsden (1978)]. The use of symplectic geometry in the study of classical mechanics has permitted to connect the Hamilton-Jacobi theory with the theory of Lagrangian submanifolds and generating functions [Barbero-Liñan, de León and Martín de Diego (2013)].

At the beginning of the 1900s an analog of Hamilton-Jacobi equation for field theory has been developed [Rund (1973)], but it has not been proved to be so powerful as the theory which is available for mechanics [Bertin, Pimentel and Pompeia (2008); Bruno (2007); Paufler and Römer (2002,b); Rosen (1971); Vitagliano (2010)].

Our goal in this chapter is to describe this equation in a geometrical setting given by the k-symplectic geometry, that is, to extend the Hamilton-Jacobi theory to field theories just in the context of k-symplectic manifolds (we remit to [de León, Marín and Marrero (1996); de León, Marrero and Martín de Diego (2009)] for a description in the multisymplectic setting). The dynamics for a given Hamiltonian function H is interpreted as a family of vector fields (a k-vector field) on the phase space $(T^1_k)^*Q$. The ***Hamilton-Jacobi equation*** is of the form

$$d(H \circ \gamma) = 0,$$

where $\gamma = (\gamma_1, \dots, \gamma_k)$ is a family of closed 1-forms on Q. Therefore, we recover the classical form

$$H\Big(q^i, \frac{\partial W^1}{\partial q^i}, \dots, \frac{\partial W^k}{\partial q^i}\Big) = constant\,,$$

where $\gamma_i = dW_i$. It should be noticed that our method is inspired in a recent result by Cariñena *et al.* [Cariñena, Gràcia, Marmo, Martínez, Muñoz-Lecanda and Román-Roy (2006)] for classical mechanics (this method has also been used to develop a Hamilton-Jacobi theory for nonholonomic mechanical systems [León, Iglesias-Ponte and Martín de Diego (2008)]; see also [Cariñena, Gràcia, Marmo, Martínez, Muñoz-Lecanda and Román-Roy (2010); de León, Marrero and Martín de Diego (2010)]).

5.1 The Hamilton-Jacobi equation

The standard formulation of the *Hamilton-Jacobi problem for Hamiltonian mechanics* consists of finding a function $S(t, q^i)$ (called the ***principal function***) such that

$$\frac{\partial S}{\partial t} + H\Big(q^i, \frac{\partial S}{\partial q^j}\Big) = 0\,. \tag{5.1}$$

If we put $S(t, q^i) = W(q^i) - t \cdot constant$, then $W\colon Q \to \mathbb{R}$ (called the ***characteristic function***) satisfies

$$H\Big(q^i, \frac{\partial W}{\partial q^j}\Big) = constant\,. \tag{5.2}$$

Equations (5.1) and (5.2) are indistinctly referred as the ***Hamilton-Jacobi equation*** in Hamiltonian mechanics.

In the framework of the k-symplectic description of classical field theory, a Hamiltonian is a function $H \in \mathcal{C}^\infty((T^1_k)^*Q)$. In this context, the Hamilton-Jacobi problem consists of finding k functions $W^1, \dots, W^k\colon Q \to \mathbb{R}$ such that

$$H\Big(q^i, \frac{\partial W^1}{\partial q^i}, \dots, \frac{\partial W^k}{\partial q^i}\Big) = constant\ . \tag{5.3}$$

In this subsection we give a geometric version of the Hamilton-Jacobi equation (5.3).

Let $\gamma : Q \longrightarrow (T^1_k)^*Q$ be a closed section of $\pi^k : (T^1_k)^*Q \longrightarrow Q$. Therefore, $\gamma = (\gamma^1, \dots, \gamma^k)$ where each γ^α is an ordinary closed 1-form on Q. Thus we have that every point has an open neighborhood $U \subset Q$ where there exist k functions $W^\alpha \in \mathcal{C}^\infty(U)$ such that $\gamma^\alpha = dW^\alpha$.

Now, let Z be a k-vector field on $(T^1_k)^*Q$. Using γ, we can construct a k-vector field Z^γ on Q such that the following diagram is commutative

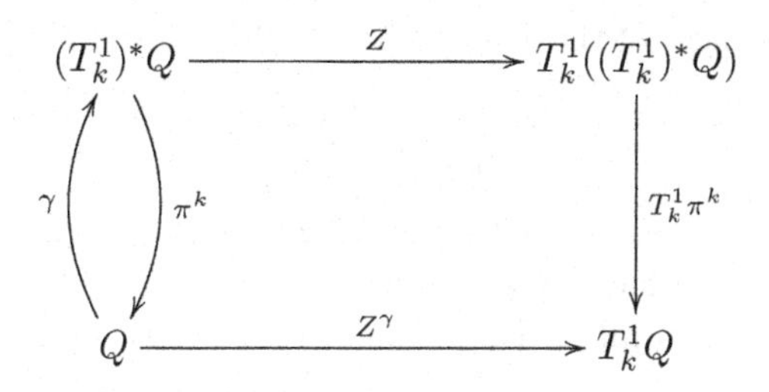

that is,

$$Z^\gamma := T^1_k\pi^k \circ Z \circ \gamma\,.$$

Let us remember that for an arbitrary differentiable map $f : M \to N$, the induced map $T^1_k f : T^1_k M \to T^1_k N$ is defined by

$$T^1_k f(v_{1x}, \dots, v_{kx}) = (f_*(x)(v_{1x}), \dots, f_*(x)(v_{kx}))\,, \tag{5.4}$$

where $v_{1x}, \dots, v_{kx} \in T_xM$, $x \in M$ and $f_*(x)\colon T_xM \to T_{f(x)}N$ is the tangent map to f at the point x

Notice that the k-vector field Z defines k vector fields on $(T^1_k)^*Q$, say $Z = (Z_1, \dots, Z_k)$. In the same way, the k-vector field Z^γ determines k vector fields on Q, say $Z^\gamma = (Z^\gamma_1, \dots, Z^\gamma_k)$.

In local coordinates, if each Z_α is locally given by

$$Z_\alpha = Z^i_\alpha \frac{\partial}{\partial q^i} + (Z_\alpha)^\beta_i \frac{\partial}{\partial p^\beta_i}\,,$$

then Z^γ_α has the following local expression:

$$Z^\gamma_\alpha = (Z^i_\alpha \circ \gamma)\, \frac{\partial}{\partial q^i}\,. \tag{5.5}$$

Let us observe that if Z is integrable, the k-vector field Z^γ is integrable.

Theorem 5.1 (Hamilton-Jacobi Theorem). *Let Z be a solution of the k-symplectic Hamiltonian equation (3.6) and $\gamma : Q \longrightarrow (T^1_k)^*Q$ be a closed section of $\pi^k : (T^1_k)^*Q \longrightarrow Q$, that is, $\gamma = (\gamma^1, \dots, \gamma^k)$ where each γ^α is an ordinary closed 1-form on Q. If Z is integrable then the following statements are equivalent:*

(1) *If $\sigma\colon U \subset \mathbb{R}^k \to Q$ is an integral section of Z^γ then $\gamma \circ \sigma$ is a solution of the Hamilton-De Donder-Weyl field equations (4.1);*

(2) $d(H \circ \gamma) = 0$.

Proof. The closeness of the 1-forms $\gamma^\alpha = \gamma_i^\alpha dq^i$ states that

$$\frac{\partial \gamma_i^\beta}{\partial q^j} = \frac{\partial \gamma_j^\beta}{\partial q^i} \,. \tag{5.6}$$

In the first place we assume that item **(1)** holds, and then we shall check that $d(H \circ \gamma) = 0$. In fact, let us suppose that $\gamma \circ \sigma(x) = (\sigma^i(x), \gamma_i^\alpha(\sigma(x)))$ is a solution of the Hamilton-De Donder-Weyl equations for H, then

$$\left.\frac{\partial \sigma^i}{\partial x^\alpha}\right|_x = \left.\frac{\partial H}{\partial p_i^\alpha}\right|_{\gamma(\sigma(x))} \quad \text{and} \quad \sum_{\alpha=1}^k \left.\frac{\partial(\gamma_i^\alpha \circ \sigma)}{\partial x^\alpha}\right|_x = -\left.\frac{\partial H}{\partial q^i}\right|_{\gamma(\sigma(x))} . \tag{5.7}$$

Now, we shall compute the differential of the function $H \circ \gamma \colon Q \to \mathbb{R}$:

$$d(H \circ \gamma) = \left(\frac{\partial H}{\partial q^i} \circ \gamma + (\frac{\partial H}{\partial p_j^\alpha} \circ \gamma) \frac{\partial \gamma_j^\alpha}{\partial q^i} \right) dq^i \,. \tag{5.8}$$

Then from (5.6), (5.7) and (5.8) we obtain

$$\begin{aligned} & d(H \circ \gamma)(\sigma(x)) \\ &= \left(\left.\frac{\partial H}{\partial q^i}\right|_{\gamma(\sigma(x))} + \left.\frac{\partial H}{\partial p_j^\alpha}\right|_{\gamma(\sigma(x))} \left.\frac{\partial \gamma_j^\alpha}{\partial q^i}\right|_{\sigma(x)} \right) dq^i(\sigma(x)) \\ &= \left(-\sum_{\alpha=1}^k \left.\frac{\partial(\gamma_i^\alpha \circ \sigma)}{\partial x^\alpha}\right|_x + \left.\frac{\partial \sigma^j}{\partial x^\alpha}\right|_x \left.\frac{\partial \gamma_j^\alpha}{\partial q^i}\right|_{\sigma(x)} \right) dq^i(\sigma(x)) \\ &= \left(-\sum_{\alpha=1}^k \left.\frac{\partial(\gamma_i^\alpha \circ \sigma)}{\partial x^\alpha}\right|_x + \left.\frac{\partial \sigma^j}{\partial x^\alpha}\right|_x \left.\frac{\partial \gamma_i^\alpha}{\partial q^j}\right|_{\sigma(x)} \right) dq^i(\sigma(x)) = 0 \,, \end{aligned}$$

the last term being zero by the chain rule. Since Z is integrable, the k-vector field Z^γ is integrable, then for each point $q \in Q$ we have an integral section $\sigma \colon U_0 \subset \mathbb{R}^k \to Q$ of Z^γ passing through this point, then

$$d(H \circ \gamma) = 0 \,.$$

Conversely, let us suppose that $d(H \circ \gamma) = 0$ and σ is an integral section of Z^γ. Now we shall prove that $\gamma \circ \sigma$ is a solution of the Hamilton field equations, that is (5.7) is satisfied.

Since $d(H \circ \gamma) = 0$, from (5.8) we obtain

$$0 = \frac{\partial H}{\partial q^i} \circ \gamma + (\frac{\partial H}{\partial p_j^\alpha} \circ \gamma) \frac{\partial \gamma_j^\alpha}{\partial q^i} \,. \tag{5.9}$$

From (3.7) and (5.5) we know that

$$Z_\alpha^\gamma = (\frac{\partial H}{\partial p_i^\alpha} \circ \gamma) \frac{\partial}{\partial q^i}$$

and then since σ is an integral section of Z^γ we obtain

$$\frac{\partial \sigma^i}{\partial x^\alpha} = \frac{\partial H}{\partial p_i^\alpha} \circ \gamma \circ \sigma \,. \tag{5.10}$$

On the other hand, from (5.6), (5.9) and (5.10) we obtain

$$\begin{aligned}\sum_{\alpha=1}^{k} \frac{\partial(\gamma_i^\alpha \circ \sigma)}{\partial x^\alpha} &= \sum_{\alpha=1}^{k} (\frac{\partial \gamma_i^\alpha}{\partial q^j} \circ \sigma) \frac{\partial \sigma^j}{\partial x^\alpha} = \sum_{\alpha=1}^{k} (\frac{\partial \gamma_i^\alpha}{\partial q^j} \circ \sigma)(\frac{\partial H}{\partial p_j^\alpha} \circ \gamma \circ \sigma) \\ &= \sum_{\alpha=1}^{k} (\frac{\partial \gamma_j^\alpha}{\partial q^i} \circ \sigma)(\frac{\partial H}{\partial p_j^\alpha} \circ \gamma \circ \sigma) = -\frac{\partial H}{\partial q^i} \circ \gamma \circ \sigma\end{aligned}$$

and thus we have proved that $\gamma \circ \sigma$ is a solution of the Hamilton-De Donder-Weyl equations. □

Remark 5.1. In the particular case $k = 1$ we reobtain the theorem proved in [de León, Marín and Marrero (1996); de León, Marrero and Martín de Diego (2009)]. ◇

Theorem 5.2. *Let Z be a solution of the k-symplectic Hamiltonian equations (3.6) and $\gamma : Q \longrightarrow (T_k^1)^*Q$ be a closed section of $\pi^k : (T_k^1)^*Q \longrightarrow Q$, that is, $\gamma = (\gamma^1, \ldots, \gamma^k)$ where each γ^α is an ordinary closed 1-form on Q. Then, the following statements are equivalent:*

(1) $Z|_{Im\gamma} - T_k^1\gamma(Z^\gamma) \in \ker \flat_\omega$, *being $\flat_\omega$ the map defined in (3.4).*
(2) $d(H \circ \gamma) = 0$.

Proof. We know that if Z_α and γ^α are locally given by

$$Z_\alpha = Z_\alpha^i \frac{\partial}{\partial q^i} + (Z_\alpha)_i^\beta \frac{\partial}{\partial p_i^\beta} \,, \quad \gamma^\alpha = \gamma_i^\alpha dq^i \,,$$

then $Z_\alpha^\gamma = (Z_\alpha^i \circ \gamma)\dfrac{\partial}{\partial q^i}$. Thus a direct computation shows that $Z|_{Im\gamma} - T_k^1\gamma(Z^\gamma) \in \ker \flat_\omega$ is locally written as

$$\left((Z_\alpha)_i^\beta \circ \gamma - (Z_A^j \circ \gamma)\frac{\partial \gamma_i^\beta}{\partial q^j}\right)\left(\frac{\partial}{\partial p_i^\beta} \circ \gamma\right) = (Y_\alpha)_i^\beta \circ \gamma \left(\frac{\partial}{\partial p_i^\beta} \circ \gamma\right) \,, \tag{5.11}$$

where $\displaystyle\sum_{\alpha=1}^{k} (Y_\alpha)_i^\alpha = 0$.

Now, we are ready to prove the result.

Assume that (**1**) holds, then from (3.7), (3.13) and (5.11) we obtain that

$$0 = \sum_{\alpha=1}^{k} \left((Z_\alpha)_i^\alpha \circ \gamma - (Z_A^j \circ \gamma) \frac{\partial \gamma_i^\alpha}{\partial q^j} \right)$$

$$= - \left((\frac{\partial H}{\partial q^i} \circ \gamma) + (\frac{\partial H}{\partial p_j^\alpha} \circ \gamma) \frac{\partial \gamma_i^\alpha}{\partial q^j} \right)$$

$$= - \left((\frac{\partial H}{\partial q^i} \circ \gamma) + (\frac{\partial H}{\partial p_j^\alpha} \circ \gamma) \frac{\partial \gamma_j^\alpha}{\partial q^i} \right)$$

where in the last identity we are using the closeness of γ (see (5.6)). Therefore, $d(H \circ \gamma) = 0$ (see (5.8)).

The converse is proved in a similar way by reversing the arguments. □

Remark 5.2. It should be noticed that if Z and Z^γ are γ-related, that is, $Z_\alpha = T\gamma(Z_\alpha^\gamma)$, then $d(H \circ \gamma) = 0$, but the converse does not hold. ◇

Corollary 5.1. *Let Z be a solution of (3.6), and γ a closed section of $\pi^k : (T_k^1)^*Q \longrightarrow Q$, as in the above theorem. If Z is integrable then the following statements are equivalent:*

(**1**) $Z_{Im\,\gamma} - T_k^1\gamma(Z^\gamma) \in \ker \flat_\omega$;

(**2**) $d(H \circ \gamma) = 0$;

(**3**) *If $\sigma\colon U \subset \mathbb{R}^k \to Q$ is an integral section of Z^γ then $\gamma \circ \sigma$ is a solution of the Hamilton-De Donder-Weyl equations.*

The equation

$$d(H \circ \gamma) = 0 \tag{5.12}$$

can be considered as the geometric version of the Hamilton-Jacobi equation for k-symplectic field theories. Notice that in local coordinates, equation (5.12) reads as

$$H(q^i, \gamma_i^\alpha(q)) = constant\,,$$

which when $\gamma^\alpha = dW^\alpha$, where $W^\alpha\colon Q \to \mathbb{R}$ is a function, takes the more familiar form

$$H(q^i, \frac{\partial W^\alpha}{\partial q^i}) = constant\,.$$

Remark 5.3. One can connect the Hamilton-Jacobi theory with the theory of Lagrangian submanifolds in the k-symplectic geometry. Let us observe that the Hamilton-Jacobi problem in the k-symplectic description consists of finding a closed section γ of π^k such that $d(H \circ \gamma) = 0$, but the condition of closed section is equivalent to finding a section γ such that its image is a k-Lagrangian submanifold of $(T^1_k)^*Q$. A proof of this equivalence and a complete description of the Lagrangian submanifolds in the k-symplectic approach can be found in [de León and Vilariño (2012)]. ◇

5.2 Example: the vibrating string problem

In this example we consider the vibrating string problem under the assumptions that the string is made up of individual particles that move vertically and $\psi(t, x)$ denotes the vertical displacement from equilibrium of the particle at horizontal position x and at time t.

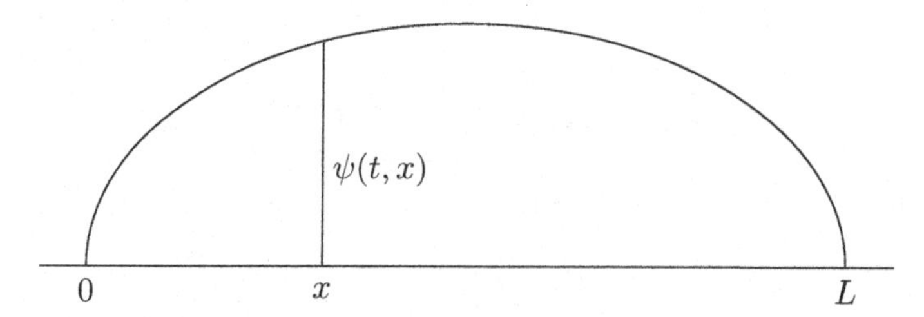

Fig. 5.1 Vibrating string at time t.

With a study of the tensile forces in this problem and using Newton's second law one obtains the equation of motion for small oscillations of a frictionless string, that is the one-dimensional wave equation

$$\sigma \frac{\partial^2 \psi}{\partial t^2} - \tau \frac{\partial^2 \psi}{\partial x^2} = 0\,, \tag{5.13}$$

where σ and τ are certain constants of the problem, σ represents the linear mass density, that is, a measure of mass per unit of length and τ is Young's module of the system related to the tension of the string, see for instance [Godstein, Poole Jr. and Safko (2001)].

Let $\gamma\colon \mathbb{R} \to (T^1_2)^*\mathbb{R}$ be the section of $\pi^2\colon T^*\mathbb{R} \oplus T^*\mathbb{R} \to \mathbb{R}$ defined by $\gamma(q) = (aq\,dq, bq\,dq)$ where a and b are two constants such that $\tau a^2 = \sigma b^2$. This section γ satisfies the condition $d(H \circ \gamma) = 0$ with H the Hamiltonian function defined by

$$\begin{aligned} H\colon\ T^*\mathbb{R}\oplus T^*\mathbb{R} &\longrightarrow \mathbb{R} \\ (q,p^1,p^2) &\to \frac{1}{2}\left(\frac{(p^1)^2}{\sigma}-\frac{(p^2)^2}{\tau}\right). \end{aligned} \tag{5.14}$$

Therefore, the condition (2) of Theorem 5.1 holds.

Let Z be a 2-vector field solution of (3.6) for the Hamiltonian (5.14), then the 2-vector field $Z^\gamma = (Z_1^\gamma, Z_2^\gamma)$ is locally given by

$$Z_1^\gamma = \frac{a}{\sigma}q\frac{\partial}{\partial q} \quad , \quad Z_2^\gamma = -\frac{b}{\tau}q\frac{\partial}{\partial q}.$$

It is easy to check that Z^γ is an integrable 2-vector field.

If $\psi\colon \mathbb{R}^2 \to \mathbb{R}$ is an integral section of Z^γ, then

$$\frac{\partial\psi}{\partial x^1} = \frac{a}{\sigma}\psi \quad , \quad \frac{\partial\psi}{\partial x^2} = -\frac{b}{\tau}\psi,$$

thus

$$\psi(x^1,x^2) = C\exp\left(\frac{a}{\sigma}x^1 - \frac{b}{\tau}x^2\right), \quad C\in\mathbb{R}.$$

By Theorem 5.1 one obtains the map $\phi = \gamma\circ\psi$, locally given by

$$(x^1,x^2) \mapsto (\psi(x^1,x^2), a\psi(x^1,x^2), b\psi(x^1,x^2)),$$

is a solution of the Hamilton-De Donder-Weyl equations associated to the Hamiltonian H given by (5.14), that is,

$$\begin{aligned} 0 &= a\frac{\partial\psi}{\partial x^1} + b\frac{\partial\psi}{\partial x^2} \\ \frac{a}{\sigma}\psi &= \frac{\partial\psi}{\partial x^1} \\ -\frac{b}{\tau}\psi &= \frac{\partial\psi}{\partial x^2}. \end{aligned}$$

Let us observe that from this system one obtains that ψ is a solution of the motion equation of the vibrating string (5.13).

Chapter 6

Lagrangian Classical Field Theories

The aim of this chapter is to give a geometric description of the ***Euler-Lagrange field equations***

$$\sum_{\alpha=1}^{k}\left(\frac{\partial^2 L}{\partial q^j \partial v^i_\alpha}\Big|_{\psi(x)}\frac{\partial \psi^j}{\partial x^\alpha}\Big|_x + \frac{\partial^2 L}{\partial v^j_\beta \partial v^i_\alpha}\Big|_{\psi(x)}\frac{\partial^2 \psi^j}{\partial x^\alpha \partial x^\beta}\Big|_x\right) = \frac{\partial L}{\partial q^i}\Big|_{\psi(x)}, \quad (6.1)$$

$1 \leq i \leq n$, where $\psi\colon \mathbb{R}^k \to T^1_kQ$ and the Lagrangian function is a function $L\colon T^1_kQ \to \mathbb{R}$ defined on the tangent bundle of k^1-velocities T^1_kQ of an arbitrary manifold Q.

Let us observe that the above equations can be written in an equivalent way as follows:

$$\sum_{\alpha=1}^{k}\frac{\partial}{\partial x^\alpha}\Big|_x\left(\frac{\partial L}{\partial v^i_\alpha}\Big|_{\psi(x)}\right) = \frac{\partial L}{\partial q^i}\Big|_{\psi(x)}, \qquad v^i_\alpha(\psi(x)) = \frac{\partial \psi^i}{\partial x^\alpha}\Big|_x. \quad (6.2)$$

The aim of this chapter is to obtain these equations in two alternative ways. Firstly, in the classical way, describing a variational principle which provides the Euler-Lagrange field equations. The second way to obtain these equations is using the k-symplectic formalism introduced in section 2.2.2.

Firstly, we shall give a detail description of T^1_kQ, i.e., the tangent bundle of k^1-velocities and we introduce some canonical geometric elements defined on this manifold. Finally we discuss the equivalence between the Hamiltonian and Lagrangian approaches when the Lagrangian function is regular or hyper-regular.

6.1 The tangent bundle of k^1-velocities

In this section we consider again (this space was introduced in section 3.1) the space T^1_kQ associated to a differentiable manifold Q and we shall give a complete description.

Each coordinate system $(q^1, \dots, q^n)$ defined on an open neighborhood $U \subset Q$, induces a local bundle coordinate system (q^i, v^i_α) on $T^1_k U \equiv (\tau^k)^{-1}(U) \subset T^1_k Q$ defined as follows

$$q^i(\mathrm{v}_q) = q^i(q), \quad v^i_\alpha(\mathrm{v}_q) = v_{\alpha q}(q^i) = (dq^i)_q(v_{\alpha q}), \tag{6.3}$$

where $\mathrm{v}_q = (v_{1q}, \dots, v_{kq}) \in T^1_k Q$, $1 \le i \le n$ and $1 \le \alpha \le k$.

These coordinates are called ***canonical coordinates*** on $T^1_k Q$ and they endow to $T^1_k Q$ of a structure of differentiable manifold of dimension $n(k+1)$.

The following diagram shows the notation which we shall use along this book:

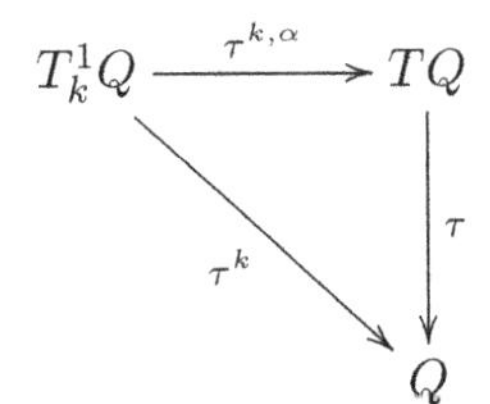

where $\tau^{k,\alpha} : T^1_k Q \to TQ$ is the canonical projection defined as follows

$$\tau^{k,\alpha}(\mathrm{v}_q) = \tau^{k,\alpha}(v_{1q}, \dots, v_{kq}) = v_{\alpha q}, \tag{6.4}$$

with $1 \le \alpha \le k$.

Remark 6.1. The manifold $T^1_k Q$ can be described as a manifold of jets, (see [de León and Rodrigues (1985); Saunders (1989)]).

Let $\phi : U_0 \subset \mathbb{R}^k \to Q$ and $\psi : V_0 \subset \mathbb{R}^k \to Q$ be two maps defined in an open neighborhood of $0 \in \mathbb{R}^k$, such that $\phi(0) = \psi(0) = p$. We say that ϕ and ψ are related on $0 \in \mathbb{R}^k$ if $\phi_*(0) = \psi_*(0)$, which means that the partial derivatives of ϕ and ψ coincide up to order one.

The equivalence classes determined by this relationship are called *jets of order 1*, or, simply, 1-jets with source $0 \in \mathbb{R}^k$ and the same target.

The 1-jet of a map $\phi : U_0 \subset \mathbb{R}^k \to Q$ is denoted by $j^1_{0,q}\phi$ where $\phi(0) = q$. The set of all 1-jets at 0 is denoted by

$$J^1_0(\mathbb{R}^k, Q) = \bigcup_{q \in Q} J^1_{0,q}(\mathbb{R}^k, Q) = \bigcup_{q \in Q} \{ j^1_{0,q}\phi \,|\, \phi \colon \mathbb{R}^k \to Q \text{ smooth}, \phi(0) = q \} .$$

The canonical projection $\beta : J^1_0(\mathbb{R}^k, Q) \to Q$ is defined by $\beta(j^1_0\phi) = \phi(0)$ and $J^1_0(\mathbb{R}^k, Q)$ is called the **tangent bundle of k^1-velocities**, (see Ehresmann [Ehresmann (1951)]). Let us observe that for $k = 1$, $J^1_0(\mathbb{R}, Q)$ is diffeomorphic to TQ.

We shall now describe the local coordinates on $J^1_0(\mathbb{R}^k, Q)$. Let U be a chart of Q with local coordinates (q^i), $1 \leq i \leq n$, $\phi : U_0 \subset \mathbb{R}^k \to Q$ a mapping such that $\phi(0) \in U$ and $\phi^i = q^i \circ \phi$. Then the 1-jet $j^1_{0,q}\phi$ is uniquely represented in $\beta^{-1}(U)$ by

$$(q^i, v^i_1, \dots, v^i_k)\,, \quad 1 \leq i \leq n$$

where

$$q^i(j^1_{0,q}\phi) = q^i(\phi(0)) = \phi^i(0)\,,\; v^i_\alpha(j^1_{0,q}\phi) = \phi_*(0)\left(\frac{\partial}{\partial x^\alpha}\Big|_0\right)(q^i)\,. \tag{6.5}$$

The manifolds T^1_kQ and $J^1_0(\mathbb{R}^k, Q)$ can be identified, via the diffeomorphism

$$\begin{array}{rcl} J^1_0(\mathbb{R}^k, Q) & \equiv & TQ \oplus \overset{k}{\dots} \oplus TQ \\ j^1_{0,q}\phi & \equiv & (v_{1q}, \dots, v_{kq}) \end{array}$$

defined by

$$v_{\alpha q} = \phi_*(0)\Big(\frac{\partial}{\partial x^\alpha}\Big|_0\Big), \quad \alpha = 1, \dots, k\,,$$

being $\phi(0) = q$. ◇

6.1.1 *Geometric elements*

In this section we introduce some geometric constructions which are necessary to describe Lagrangian classical field theories using the k-symplectic approach.

Vertical lifts

Given a tangent vector u_q on an arbitrary manifold Q, one can consider the vertical lift to the tangent bundle of TQ. In a similar way, we can define the vertical lift to the tangent bundle of k^1-velocities by considering the lift on each copy of the tangent bundle.

Definition 6.1. Let $u_q \in T_qQ$ be a tangent vector at $q \in Q$. For each $1 \leq \alpha \leq k$, we define the ***vertical α-lift***, $(u_q)^{V_\alpha}$, as the vector field at the fiber $(\tau^k)^{-1}(q) \subset T^1_kQ$ given by

$$(u_q)^{V_\alpha}_{\mathrm{v}_q} = \frac{d}{ds}(v_{1q}, \dots, v_{\alpha-1\,q}, v_{\alpha q} + s u_q, v_{\alpha+1\,q}, \dots, v_{kq})\Big|_{s=0} \tag{6.6}$$

for any point $\mathrm{v}_q = (v_{1q}, \dots, v_{kq}) \in (\tau^k)^{-1}(q) \subset T^1_kQ$.

In local canonical coordinates (6.3), if $u_q = u^i \left.\dfrac{\partial}{\partial q^i}\right|_q$ then

$$(u_q)^{V_\alpha}_{\mathrm{v}_q} = u^i \left.\frac{\partial}{\partial v^i_\alpha}\right|_{\mathrm{v}_q}. \tag{6.7}$$

The vertical lifts of tangent vectors allow us to define the vertical lift of vector fields.

Definition 6.2. Let X be a vector field on Q. For each $1 \leq \alpha \leq k$ we call ***α-vertical lift*** of X to T^1_kQ, to the vector field $X^{V_\alpha} \in \mathfrak{X}(T^1_kQ)$ defined by

$$X^{V_\alpha}(\mathrm{v}_q) = (X(q))^{V_\alpha}_{\mathrm{v}_q}, \tag{6.8}$$

for all points $\mathrm{v}_q = (v_{1q}, \dots, v_{kq}) \in T^1_kQ$.

If $X = X^i \dfrac{\partial}{\partial q^i}$ then, from (6.7) and (6.8) we deduce that

$$X^{V_\alpha} = (X^i \circ \tau^k) \frac{\partial}{\partial v^i_\alpha}, \tag{6.9}$$

since

$$(X(q))^{V_\alpha}_{\mathrm{v}_q} = \left(X^i(q) \left.\frac{\partial}{\partial q^i}\right|_q \right)^{V_\alpha}_{\mathrm{v}_q} = X^i(q) \left.\frac{\partial}{\partial v^i_\alpha}\right|_{\mathrm{v}_q} = (X^i \circ \tau^k)(\mathrm{v}_q) \left.\frac{\partial}{\partial v^i_\alpha}\right|_{\mathrm{v}_q}.$$

Canonical k-tangent structure

In a similar way as in the tangent bundle, the vertical lifts of tangent vectors allows us to introduce a family $\{J^1, \dots, J^k\}$ of k tensor fields of type $(1,1)$ on T^1_kQ. This family is the model of the so-called ***k-tangent structures*** introduced by M. de León *et al.* in [de León, Méndez and Salgado (1988, 1991)]. In the case $k = 1$, $J = J^1$ is the canonical tangent structure or vertical endomorphism (1.13) (see [Crampin (1983,b); Grifone (1972,b); Grifone and Mehdi (1999); Klein (1962)]).

Definition 6.3. For each $1 \leq \alpha \leq k$ we define the tensor field J^α of type $(1,1)$ on T^1_kQ as follows

$$\begin{aligned} J^\alpha(\mathrm{v}_q) : T_{\mathrm{v}_q}(T^1_kQ) &\to T_{\mathrm{v}_q}(T^1_kQ) \\ Z_{\mathrm{v}_q} &\to J^\alpha(\mathrm{v}_q)(Z_{\mathrm{v}_q}) = \Big((\tau^k)_*(\mathrm{v}_q)(Z_{\mathrm{v}_q}) \Big)^{V_\alpha}_{\mathrm{v}_q} \end{aligned} \tag{6.10}$$

where $\mathrm{v}_q \in T^1_kQ$.

From (6.7) and (6.10) we deduce that, for each $1 \leq \alpha \leq k$, J^α is locally given by

$$J^\alpha = \frac{\partial}{\partial v^i_\alpha} \otimes dq^i \,. \tag{6.11}$$

Remark 6.2. The family $\{J^1, \dots, J^k\}$ can be obtained using the vertical lifts of the identity tensor field of Q to T^1_kQ defined by Morimoto (see [Morimoto (1969, 1970)]). ◇

Canonical vector fields

An important geometric object on T^1_kQ is the generalized Liouville vector field.

Definition 6.4. The **Liouville vector field** $\triangle$ on T^1_kQ is the infinitesimal generator of the flow

$$\begin{aligned} \psi : \mathbb{R} \times T^1_kQ &\longrightarrow T^1_kQ \\ (s, (v_{1_q}, \dots, v_{k_q})) &\mapsto (e^s v_{1_q}, \dots, e^s v_{k_q}) \end{aligned} \tag{6.12}$$

and in local coordinates it has the form

$$\triangle = \sum_{i=1}^n \sum_{\alpha=1}^k v^i_\alpha \frac{\partial}{\partial v^i_\alpha} \,. \tag{6.13}$$

Definition 6.5. For each $1 \leq \alpha \leq k$ we define the *canonical vector field* $\triangle_\alpha$ as the infinitesimal generator of the following flow

$$\begin{aligned} \psi^\alpha : \mathbb{R} \times T^1_kQ &\longrightarrow T^1_kQ \\ (s, (v_{1_q}, \dots, v_{k_q})) &\mapsto (v_{1_q}, \dots, v_{\alpha-1_q}, e^s v_{\alpha_q}, v_{\alpha+1_q}, \dots, v_{k_q}) \,, \end{aligned} \tag{6.14}$$

and in local coordinates it has the form

$$\triangle_\alpha = \sum_{i=1}^n v^i_\alpha \frac{\partial}{\partial v^i_\alpha} \,, \quad 1 \leq \alpha \leq k \,, \tag{6.15}$$

for each $1 \leq \alpha \leq k$.

From (6.13) and (6.15) we deduce that $\triangle = \triangle_1 + \ldots + \triangle_k$.

Remark 6.3. The vector fields $\triangle$ and $\triangle_\alpha$ can also be defined using the vertical lifts. From (6.7), (6.13) and (6.15) one obtains that

$$\triangle(\mathrm{v}_q) = \sum_{\alpha=1}^k (v_{\alpha_q})^{V_\alpha}_{\mathrm{v}_q} \,, \quad \triangle_\alpha(\mathrm{v}_q) = (v_{\alpha_q})^{V_\alpha}_{\mathrm{v}_q} \,,$$

where $\mathrm{v}_q = (v_{1_q}, \dots, v_{k_q}) \in T^1_kQ$ and $1 \leq \alpha \leq k$. ◇

6.1.2 Prolongation of vector fields

In a similar way as in section 4.1.1 one can define the canonical prolongation of maps between manifolds to the corresponding tangent bundles of k^1-velocities (see (5.4)).

Definition 6.6. Let $Z \in \mathfrak{X}(Q)$ be a vector field on Q with local 1-parametric group of diffeomorphisms $h_s : Q \to Q$. The **complete or natural lift** of Z to T^1_kQ is the vector field Z^C on T^1_kQ whose local 1-parameter group of diffeomorphisms is $T^1_k(h_s) : T^1_kQ \to T^1_kQ$.

Remark 6.4. The definition of $T^1_k(h_s)$ is just the one gives in (5.4). ◇

In local canonical coordinates (6.3), if $Z = Z^i \dfrac{\partial}{\partial q^i}$ then the local expression is

$$Z^C = Z^i \frac{\partial}{\partial q^i} + v^j_\alpha \frac{\partial Z^i}{\partial q^j} \frac{\partial}{\partial v^i_\alpha} \,. \tag{6.16}$$

The following lemma shows that the canonical prolongations of maps to the tangent bundle of k^1-velocities leave invariant the canonical structures of T^1_kQ.

Lemma 6.1. *Let $\Phi = T^1_k\varphi : T^1_kQ \to T^1_kQ$ be the canonical prolongation of a diffeomorphism $\varphi : Q \to Q$, then for each $1 \leq \alpha \leq k$, we have*

$$(a) \quad \Phi_* \circ J^\alpha = J^\alpha \circ \Phi_* \,, \quad (b) \quad \Phi_* \Delta_\alpha = \Delta_\alpha \,, \quad (c) \quad \Phi_* \Delta = \Delta \,.$$

Proof. (a) It is a direct consequence of the local expression (6.11) of J^α and the local expression of $T^1_k\varphi$ given by

$$T^1_k\varphi(q^i, v^i_\alpha) = \left(\varphi^j(q^i), v^i_\alpha \frac{\partial \varphi^j}{\partial q^i}\right)$$

where the functions φ^j denote the components of the diffeomorphism $\varphi : Q \to Q$.

(b) It is a consequence of $T^1_k\varphi \circ \psi^\alpha_{\mathrm{v}_q} = \psi^\alpha_{\mathrm{v}_q} \circ T^1_k\varphi$, where $\psi^\alpha_{\mathrm{v}_q}$ are 1-parameter group of diffeomorphisms (6.14) generated by Δ_α.

(c) It is a direct consequence of (b) and of the identity $\Delta = \Delta_1 + \ldots + \Delta_k$. □

6.1.3 *First prolongation of maps*

Here we shall introduce the notion of first prolongation, which will be very important in this chapter and generalize the lift of a curve on Q to the tangent bundle TQ of Q.

Definition 6.7. We define the ***first prolongation*** $\phi^{(1)}$ of a map $\phi\colon \mathbb{R}^k \to Q$ as the map

$$\begin{aligned} \phi^{(1)}\colon\ U_0 \subseteq \mathbb{R}^k &\longrightarrow T^1_kQ \\ x &\longmapsto \Big(\phi_*(x)\Big(\frac{\partial}{\partial x^1}\Big|_x\Big), \dots, \phi_*(x)\Big(\frac{\partial}{\partial x^k}\Big|_x\Big)\Big), \end{aligned} \tag{6.17}$$

where $(x^1, \dots, x^k)$ denotes the coordinates on $\mathbb{R}^k$ and T^1_kQ the tangent bundle of k^1-velocities introduced at the beginning of section 6.1.

If we consider canonical coordinates (q^i, v^i_α) on T^1_kQ (see (6.3) for the definition), then the first prolongation is locally given by

$$\begin{aligned} \phi^{(1)}\colon\ U_0 \subseteq \mathbb{R}^k &\longrightarrow T^1_kQ \\ x &\longmapsto \phi^{(1)}(x) = \Big(\phi^i(x), \frac{\partial \phi^i}{\partial x^\alpha}\Big|_x\Big), \end{aligned} \tag{6.18}$$

where $\phi^i = q^i \circ \phi$, and we are using that

$$\phi_*(x)\Big(\frac{\partial}{\partial x^\alpha}\Big|_x\Big) = \frac{\partial \phi^i}{\partial x^\alpha}\Big|_x \frac{\partial}{\partial q^i}\Big|_{\phi(x)}.$$

6.2 Variational principle for the Euler-Lagrange equations

In this section we describe the problem in the calculus of variations for multiple integrals, which allows us to obtain the Euler-Lagrange field equations.

Along this section we consider a given Lagrangian function L on the tangent bundle of k^1-velocities, i.e., $L\colon T^1_kQ \to \mathbb{R}$. Thus we can evaluate L in the first prolongation (6.17) of a field $\phi\colon \mathbb{R}^k \to Q$. Given L we can construct the following operator:

Definition 6.8. Let us denote by $\mathcal{C}^\infty_C(\mathbb{R}^k, Q)$ the set of maps $\phi : U_0 \subset \mathbb{R}^k \to Q$, with compact support, defined on an open set U_0. We define the action associated to L by

$$\begin{aligned} \mathcal{J} : \mathcal{C}^\infty_C(\mathbb{R}^k, Q) &\to \mathbb{R} \\ \phi &\mapsto \mathcal{J}(\phi) = \int_{\mathbb{R}^k} (L \circ \phi^{(1)})(x)\, d^kx\,, \end{aligned}$$

where $d^kx = dx^1 \wedge \ldots \wedge dx^k$ is a volume form on $\mathbb{R}^k$ and $\phi^{(1)} : U_0 \subset \mathbb{R}^k \to T^1_kQ$ denotes the first prolongation of ϕ defined in (6.17).

Definition 6.9. A map $\phi \in \mathcal{C}^\infty_C(\mathbb{R}^k, Q)$, is an *extremal* of $\mathcal{J}$ if

$$\frac{d}{ds}\Big|_{s=0} \mathcal{J}(\tau_s \circ \phi) = 0,$$

for each flow τ_s on Q such that $\tau_s(q) = q$ for every q at the boundary of $\phi(U_0) \subset Q$.

Let us observe that the flow $\tau_s : Q \to Q$, considered in this definition, are generated by a vector field on Q which vanishes at the boundary of $\phi(U_0)$.

The variational problem associated to a Lagrangian L, is to find the extremals of the integral action $\mathcal{J}$. In the following proposition we characterize these extremals.

Prop 6.1. Let $L : T^1_kQ \to \mathbb{R}$ be a Lagrangian and $\phi \in \mathcal{C}^\infty_C(\mathbb{R}^k, Q)$. The following assertions are equivalent :

(1) $\phi : U_0 \subset \mathbb{R}^k \to Q$ is an extremal of $\mathcal{J}$.
(2) For each vector field Z on Q, vanishing at all points on the boundary of $\phi(U_0)$, one has

$$\int_{U_0} \left((\mathcal{L}_{Z^C} L) \circ \phi^{(1)}\right)(x) d^kx = 0\,,$$

where Z^C is the complete lift of Z to T^1_kQ (see (6.6)).
(3) ϕ is a solution of the Euler-Lagrange field equations (6.2).

Proof. First we prove the equivalence between (1) and (2).

Let $\phi\colon U_0 \subset \mathbb{R}^k \to Q$ be a map and $Z \in \mathfrak{X}(Q)$ be a vector field on Q, with local 1-parameter group of diffeomorphism $\{\tau_s\}$, and vanishing at the boundary of $\phi(U_0)$, then $T^1_k\tau_s$ is the local 1-parameter group of diffeomorphism of Z^C.

A simple computation shows $T^1_k\tau_s \circ \phi^{(1)} = (\tau_s \circ \phi)^{(1)}$, and thus we deduce

$$\frac{d}{ds}\Big|_{s=0} \mathcal{J}(\tau_s \circ \phi) = \frac{d}{ds}\Big|_{s=0} \int_{\mathbb{R}^k} (L \circ (\tau_s \circ \phi)^{(1)})(x) d^kx$$
$$= \lim_{s \to 0} \frac{1}{s}\left(\int_{\mathbb{R}^k} (L \circ (\tau_s \circ \phi)^{(1)})(x) d^kx - \int_{\mathbb{R}^k} (L \circ \tau_0 \circ \phi^{(1)})(x) d^kx\right)$$

$$= \lim_{s\to 0}\frac{1}{s}\left(\int_{\mathbb{R}^k}(L\circ T^1_k\tau_s\circ\phi^{(1)})(x)d^kx - \int_{\mathbb{R}^k}(L\circ\phi^{(1)})(x)d^kx\right)$$

$$= \lim_{s\to 0}\frac{1}{s}\left(\int_{\mathbb{R}^k}\Big((L(T^1_k\tau_s\circ\phi^{(1)})(x)) - L(\phi^{(1)}(x))\Big)d^kx\right)$$

$$= \int_{\mathbb{R}^k}\left(\lim_{s\to 0}\frac{1}{s}\Big(L(T^1_k\tau_s\circ\phi^{(1)}(x)) - L(\phi^{(1)}(x))\Big)\right)d^kx$$

$$= \int_{\mathbb{R}^k}\left((\mathcal{L}_{Z^c}L)\circ\phi^{(1)}\right)(x)d^kx\,,$$

so, we are done.

We have proved that $\phi : U_0\subset\mathbb{R}^k\to Q$ is an extremal of $\mathcal{J}$ if and only if for each vector field $Z\in\mathfrak{X}(Q)$ vanishing at the boundary of $\phi(U_0)$ one has

$$\int_{U_0}\left((\mathcal{L}_{Z^c}L)\circ\phi^{(1)}\right)(x)d^kx = 0\,. \tag{6.19}$$

We now prove that it is equivalent to saying that ϕ is a solution of the Euler-Lagrange field equation.

Let us suppose that $Z = Z^i\dfrac{\partial}{\partial q^i}$; from the local expression (6.16) of Z^C and the expression of integration by parts in multiple integrals and since ϕ has compact support, we deduce that:

$$\int_{\mathbb{R}^k}\left((\mathcal{L}_{Z^c}L)\circ\phi^{(1)}\right)(x)d^kx$$

$$= \int_{\mathbb{R}^k}\left(Z^i(\phi(x))\frac{\partial L}{\partial q^i}\Big|_{\phi^{(1)}(x)} + \frac{\partial\phi^j}{\partial x^\alpha}\Big|_x\frac{\partial Z^i}{\partial q^j}\Big|_{\phi(x)}\frac{\partial L}{\partial v^i_\alpha}\Big|_{\phi^{(1)}(x)}\right)d^kx$$

$$= \int_{\mathbb{R}^k}\left(Z^i(\phi(x))\frac{\partial L}{\partial q^i}\Big|_{\phi^{(1)}(x)} + \frac{\partial(Z^i\circ\phi)}{\partial x^\alpha}\Big|_x\frac{\partial L}{\partial q^i}\Big|_{\phi^{(1)}(x)}\right)d^kx$$

$$= \int_{\mathbb{R}^k}\left(Z^i(\phi(x))\frac{\partial L}{\partial q^i}\Big|_{\phi^{(1)}(x)} - Z^i(\phi(x))\frac{\partial}{\partial x^\alpha}\left(\frac{\partial L}{\partial v^i_\alpha}\Big|_{\phi^{(1)}(x)}\right)\right)d^kx$$

$$= \int_{\mathbb{R}^k}(Z^i\circ\phi)(x)\left(\frac{\partial L}{\partial q^i}\Big|_{\phi^{(1)}(x)} - \frac{\partial}{\partial x^\alpha}\left(\frac{\partial L}{\partial v^i_\alpha}\Big|_{\phi^{(1)}(x)}\right)\right)d^kx\,.$$

Therefore we obtain that ϕ is an extremal of $\mathcal{J}$ if and only if

$$0 = \int_{\mathbb{R}^k}(Z^i\circ\phi)(x)\left(\frac{\partial L}{\partial q^i}\Big|_{\phi^{(1)}(x)} - \frac{\partial}{\partial x^\alpha}\left(\frac{\partial L}{\partial v^i_\alpha}\Big|_{\phi^{(1)}(x)}\right)\right)d^kx\,.$$

Since this identity holds for all Z^i, applying Lemma 4.2 we obtain that ϕ is an extremal of $\mathcal{J}$ if and only if

$$\sum_{\alpha=1}^{k} \frac{\partial}{\partial x^\alpha}\Big|_x \left(\frac{\partial L}{\partial v^i_\alpha}\Big|_{\phi^{(1)}(x)} \right) = \frac{\partial L}{\partial q^i}\Big|_{\phi^{(1)}(x)} . \tag{6.20}$$

Equations (6.20) are called **Euler-Lagrange field equations** for the Lagrangian function L. □

6.3 Euler-Lagrange field equations: k-symplectic version

In this section we give the geometric description of the Euler-Lagrange field equations (6.2) or (6.20). In order to accomplish this task it is necessary to introduce some geometric elements associated to a Lagrangian function $L\colon T^1_kQ \to \mathbb{R}$ (see for instance [de León and Rodrigues (1985)]).

6.3.1 *Poincaré-Cartan forms on the tangent bundle of k^1-velocities*

In a similar manner as in the case of Lagrangian mechanics, the k-tangent structure on T^1_kQ, allows us to define a family of 1-forms, $\theta^1_L, \dots, \theta^k_L$ on T^1_kQ as follows:

$$\theta^\alpha_L = dL \circ J^\alpha \,, \tag{6.21}$$

where $1 \leq \alpha \leq k$. Next we define the family $\omega^1_L, \dots, \omega^k_L$ of presymplectic forms on T^1_kQ by

$$\omega^\alpha_L = -\, d\theta^\alpha_L \,, \tag{6.22}$$

which will be called ***Poincaré-Cartan forms*** on T^1_kQ.

If we consider canonical coordinates (q^i, v^i_α) on T^1_kQ, from (6.11) and (6.21) we deduce that for $1 \leq \alpha \leq k$,

$$\theta^\alpha_L = \frac{\partial L}{\partial v^i_\alpha} dq^i \,, \tag{6.23}$$

and so, from (6.22) and (6.23), we obtain

$$\omega^\alpha_L = dq^i \wedge d\left(\frac{\partial L}{\partial v^i_\alpha}\right) = \frac{\partial^2 L}{\partial q^j \partial v^i_\alpha} dq^i \wedge dq^j + \frac{\partial^2 L}{\partial v^j_\beta \partial v^i_\alpha} dq^i \wedge dv^j_\beta \,. \tag{6.24}$$

An important property of the family of presymplectic forms $\omega^1_L, \dots, \omega^k_L$ occurs when the Lagrangian is regular.

Definition 6.10. A Lagrangian function $L\colon T^1_kQ \to \mathbb{R}$ is said to be ***regular*** if the matrix

$$\left(\frac{\partial^2 L}{\partial v^i_\alpha \partial v^j_\beta}\right)$$

is regular.

The regularity condition let us prove the following proposition, see [Munteanu, Rey and Salgado (2004)].

Prop 6.2. Given a Lagrangian function on T^1_kQ, the following conditions are equivalent:

(1) L is regular.

(2) $(\omega^1_L, \dots, \omega^k_L, V)$ is a k-symplectic structure on T^1_kQ, where

$$V = \ker(\tau^k)_* = span\left\{\frac{\partial}{\partial v^i_1}, \dots, \frac{\partial}{\partial v^i_k}\right\}$$

with $1 \leq i \leq n$, is the vertical distribution of the vector bundle $\tau^k : T^1_kQ \to Q$.

6.3.2 *Second order partial differential equations on T^1_kQ*

The second geometric notion which we need in our description of the Euler-Lagrange equations is the notion of second order partial differential equation (or SOPDE) on T^1_kQ. Roughly speaking, a SOPDE is a k-vector field on T^1_kQ whose integral sections are first prolongations of maps $\phi\colon \mathbb{R}^k \to Q$.

In this section it is fundamental to recall the notion of k-vector field and integral section introduced in section 3.1. Now, we only consider k-vector fields on $M = T^1_kQ$. Thus using local coordinates (q^i, v^i_α) on an open set T^1_kU, the local expression of a k-vector field $X = (X_1, \dots, X_k)$ on T^1_kQ is given by

$$X_\alpha = (X_\alpha)^i \frac{\partial}{\partial q^i} + (X_\alpha)^i_\beta \frac{\partial}{\partial v^i_\beta}, \quad (1 \leq \alpha \leq k)\,. \tag{6.25}$$

Let

$$\varphi\colon U_0 \subset \mathbb{R}^k \to T^1_kQ$$

be an integral section of $(X_1, \dots, X_k)$ with components

$$\varphi(x) = (\psi^i(x), \psi^i_\alpha(x))\,.$$

Then since

$$\varphi_*(x)\Big(\frac{\partial}{\partial x^\alpha}\Big|_x\Big) = \frac{\partial\psi^i}{\partial x^\alpha}\Big|_x \frac{\partial}{\partial q^i}\Big|_{\varphi(x)} + \frac{\partial\psi^i_\beta}{\partial x^\alpha}\Big|_x \frac{\partial}{\partial v^i_\beta}\Big|_{\varphi(x)}$$

the condition of integral section (3.2) for this case is locally equivalent to the following system of partial differential equations (condition (3.3))

$$\frac{\partial\psi^i}{\partial x^\alpha}\Big|_x = (X_\alpha)^i(\varphi(x)), \quad \frac{\partial\psi^i_\beta}{\partial x^\alpha}\Big|_x = (X_\alpha)^i_\beta(\varphi(x)), \tag{6.26}$$

with $1 \leq i \leq n$ and $1 \leq \alpha, \beta \leq k$.

Definition 6.11. A **second order partial differential equation** (or SOPDE to short) is a k-vector field $\mathbf{X} = (X_1, \ldots, X_k)$ on T^1_kQ, which is a section of the projection $T^1_k\tau^k : T^1_k(T^1_kQ) \to T^1_kQ$, i.e.,

$$\tau^k_{T^1_kQ} \circ \mathbf{X} = id_{T^1_kQ} \text{ and } T^1_k\tau^k \circ \mathbf{X} = id_{T^1_kQ} ,$$

where $\tau^k \colon T^1_kQ \to Q$ and $\tau^k_{T^1_kQ} \colon T^1_k(T^1_kQ) \to T^1_kQ$ are the canonical projections.

Let us observe that when $k = 1$ this definition coincides with the definition of SODE (second order differential equation), see for instance [de León and Rodrigues (1985)].

Taking into account the definition of $T^1_k\tau^k$ (see Definition 5.4), the above definition is equivalent to saying that a k-vector field $(X_1, \ldots, X_k)$ on T^1_kQ is a SOPDE if and only if

$$(\tau^k)_*(\mathrm{v}_q)(X_\alpha(\mathrm{v}_q)) = v_{\alpha_q} ,$$

for $1 \leq \alpha \leq k$, where $\mathrm{v}_q = (v_{1_q}, \ldots, v_{k_q}) \in T^1_kQ$.

If we now consider the canonical coordinate system (q^i, v^i_α), from (6.25) and Definition 6.11, the local expression of a SOPDE $\mathbf{X} = (X_1, \ldots, X_k)$ is the following:

$$X_\alpha(q^i, v^i_\alpha) = v^i_\alpha \frac{\partial}{\partial q^i} + (X_\alpha)^i_\beta \frac{\partial}{\partial v^i_\beta}, \tag{6.27}$$

where $1 \leq \alpha \leq k$ and $(X_\alpha)^i_\beta$ are functions on T^1_kQ.

In the case $k = 1$, the integral curves of as SODE on TQ are lifts to TQ of curves on Q. In our case, in order to characterize the integral sections of a SOPDE we consider Definition 6.17 of the first prolongation $\phi^{(1)}$ of a map $\phi \colon \mathbb{R}^k \to Q$ to T^1_kQ.

Consider a SOPDE $\mathbf{X} = (X_1, \ldots, X_k)$ and a map

$$\begin{aligned} \varphi\colon\ \mathbb{R}^k &\to T^1_kQ \\ x &\to \varphi(x) = (\psi^i(x), \psi^i_\beta(x)) \end{aligned}$$

Since a SOPDE $\mathbf{X}$ is, in particular, a k-vector field on T_k^1Q, from (6.26) and (6.27) one obtains that φ is an integral section of $\mathbf{X}$ if and only if φ is a solution of the following system of partial differential equations:

$$\left.\frac{\partial \psi^i}{\partial x^\alpha}\right|_x = v_\alpha^i(\varphi(x)) = \psi_\alpha^i(x)\,, \qquad \left.\frac{\partial \psi_\beta^i}{\partial x^\alpha}\right|_x = (X_\alpha)_\beta^i(\varphi(x))\,, \tag{6.28}$$

with $1 \le i \le n$ and $1 \le \alpha \le k$.

Thus, from (6.18) and (6.28) it is easy to prove the following proposition.

Prop 6.3. Let $\mathbf{X} = (X_1, \dots, X_k)$ be an integrable SOPDE.

(1) If φ is an integral section of $\mathbf{X}$ then $\varphi = \phi^{(1)}$, where $\phi^{(1)} \colon \mathbb{R}^k \to T_k^1Q$ is the first prolongation of the map

$$\phi := \tau^k \circ \varphi : \mathbb{R}^k \xrightarrow{\ \varphi\ } T_k^1 Q \xrightarrow{\ \tau^k\ } Q\ .$$

Moreover, $\phi(x) = (\psi^i(x))$ is a solution of the system of second order partial differential equations

$$\left.\frac{\partial^2 \psi^i}{\partial x^\alpha \partial x^\beta}\right|_x = (X_\alpha)_\beta^i(\psi^i(x), \frac{\partial \psi^i}{\partial x^\gamma}(x))\,, \tag{6.29}$$

with $1 \le i \le n\,; 1 \le \alpha, \beta, \gamma \le k$.

(2) Conversely, if $\phi : \mathbb{R}^k \to Q$, locally given by $\phi(x) = (\psi^i(x))$, is a map satisfying (6.29) then $\phi^{(1)}$ is an integral section of $\mathbf{X} = (X_1, \dots, X_k)$.

□

Remark 6.5. From equation (6.29) we deduce that, when the SOPDE $\mathbf{X}$ is integrable (as a k-vector field), we have $(X_\alpha)_\beta^i = (X_\beta)_\alpha^i$ for all $\alpha, \beta = 1, \dots, k$ and $1 \le i \le n$. ◇

The following characterization of SOPDEs on T_k^1Q can be given using the canonical k-tangent structure $J^1, \dots, J^k$ and the canonical vector fields $\Delta_1, \dots, \Delta_k$ (these object were introduced in section 6.1.1).

Prop 6.4. Let $\mathbf{X} = (X_1, \dots, X_k)$ be a k-vector field on T_k^1Q. The following conditions are equivalent

(1) $\mathbf{X}$ is a SOPDE.
(2) $J^\alpha(X_\alpha) = \Delta_\alpha$, for all $1 \le \alpha \le k$.

Proof. It is an immediate consequence of (6.15) and (6.27). □

6.3.3 *Euler-Lagrange field equations*

In this subsection we describe the Lagrangian formulation of classical field theories using the geometrical elements introduced in the previous sections of this book.

In a similar way as in the Hamiltonian case, given a Lagrangian function $L\colon T^1_kQ \to \mathbb{R}$, we now consider the manifold T^1_kQ equipped with the Poincaré-Cartan forms $(\omega^1_L, \dots, \omega^k_L)$ defined in section 6.3.1, which allows us to define a k-symplectic structure on T^1_kQ when the Lagrangian function is regular.

Denote by $\mathfrak{X}^k_L(T^1_kQ)$ the set of k-vector fields $\mathbf{X} = (X_1, \dots, X_k)$ in T^1_kQ, which are solutions of the equation

$$\sum_{\alpha=1}^{k} \iota_{X_\alpha}\omega^\alpha_L = \mathrm{d}E_L \,, \tag{6.30}$$

where E_L is the function on T^1_kQ defined by $E_L = \Delta(L) - L$.

Consider canonical coordinates (q^i, v^i_α) on T^1_kQ, then each X_α is locally given by the expression (6.25).

Now, from (6.13) we obtain that the function E_L is locally given

$$E_L = v^i_\alpha \frac{\partial L}{\partial v^i_\alpha} - L$$

and then

$$dE_L = \left(v^i_\alpha \frac{\partial^2 L}{\partial q^j \partial v^i_\alpha} - \frac{\partial L}{\partial q^j} \right) dq^j + v^i_\alpha \frac{\partial^2 L}{\partial v^i_\alpha \partial v^j_\beta} dv^j_\beta \,. \tag{6.31}$$

Therefore, from (6.24), (6.25) and (6.31) one obtains that a k-vector field $\mathbf{X} = (X_1, \dots, X_k)$ on T^1_kQ is a solution of (6.30) if, and only if, the functions $(X_\alpha)^i$ and $(X_\alpha)^i_\beta$ satisfy the following local system of equations

$$\left(\frac{\partial^2 L}{\partial q^i \partial v^j_\alpha} - \frac{\partial^2 L}{\partial q^j \partial v^i_\alpha} \right) (X_\alpha)^j - \frac{\partial^2 L}{\partial v^i_\alpha \partial v^j_\beta} (X_\alpha)^j_\beta = v^j_\alpha \frac{\partial^2 L}{\partial q^i \partial v^j_\alpha} - \frac{\partial L}{\partial q^i} \,, \tag{6.32}$$

$$\frac{\partial^2 L}{\partial v^j_\beta \partial v^i_\alpha} (X_\alpha)^i = \frac{\partial^2 L}{\partial v^j_\beta \partial v^i_\alpha} v^i_\alpha , \tag{6.33}$$

where $1 \leq \alpha, \beta \leq k$ and $1 \leq i, j \leq n$.

If the Lagrangian is regular, the above equations are equivalent to the equations

$$\frac{\partial^2 L}{\partial q^j \partial v^i_\alpha} v^j_\alpha + \frac{\partial^2 L}{\partial v^i_\alpha \partial v^j_\beta} (X_\alpha)^j_\beta = \frac{\partial L}{\partial q^i} \,, \tag{6.34}$$

$$(X_\alpha)^i = v^i_\alpha \,, \tag{6.35}$$

where $1 \leq \alpha, \beta \leq k$ and $1 \leq i, j \leq n$.

Thus, we can state the following theorem.

Theorem 6.1. *Let $L : T^1_kQ \to \mathbb{R}$ be a Lagrangian and $\mathbf{X} = (X_1, \dots, X_k) \in \mathfrak{X}^k_L(T^1_kQ)$. Then,*

(1) *If L is regular then $\mathbf{X} = (X_1, \dots, X_k)$ is a* SOPDE.
Moreover if $\varphi : \mathbb{R}^k \to T^1_kQ$ is an integral section of $\mathbf{X}$, then the map $\phi = \tau^k \circ \varphi \colon \mathbb{R}^k \to Q$ is a solution of the Euler-Lagrange field equations (6.20).

(2) *If $\mathbf{X} = (X_1, \dots, X_k)$ is integrable and $\phi^{(1)} : \mathbb{R}^k \to T^1_kQ$ is an integral section of $\mathbf{X}$ then $\phi : \mathbb{R}^k \to Q$ is a solution of the Euler-Lagrange field equations (6.20).*

Proof. **(1)** Let L be a regular Lagrangian, then $\mathbf{X} = (X_1, \dots, X_k) \in \mathfrak{X}^k_L(T^1_kQ)$ if the coefficients of X satisfy (6.34) and (6.35). The expression (6.35) is locally equivalent to saying that $\mathbf{X}$ is a SOPDE.
Since in this case $\mathbf{X}$ is a SOPDE, we can apply Proposition 6.3, therefore, if $\varphi \colon \mathbb{R}^k \to T^1_kQ$ is an integral section of X, then $\varphi = \phi^{(1)}$.
Finally, from (6.28) and (6.34) one obtains that ϕ is a solution of the Euler-Lagrange equations (6.2).

(2) In this case we suppose that $\phi^{(1)}$ is an integral section of $\mathbf{X}$, then in a similar way as in Proposition 6.3(1), one can prove that the components ϕ^i, with $1 \leq i \leq n$, of ϕ satisfy (6.29).
Thus from (6.28), (6.29), (6.32) and (6.33) one obtains that ϕ is a solution of the Euler-Lagrange equations (6.2).

□

Remark 6.6. If we write an equation (6.30) for the case $k = 1$, we obtain

$$\iota_X \omega_L = dE_L$$

which is the equation of the geometric formulation of the Lagrangian mechanics in symplectic terms. ◇

Remark 6.7. One important difference with the case $k = 1$ on the tangent bundle TQ is that for an arbitrary k we cannot ensure the unicity of solutions of equation (6.30).

When the Lagrangian L is regular, Proposition 6.2 implies that $(T^1_kQ, \omega^1_L, \dots, \omega^k_L, V)$ is a k-symplectic manifold and equation (6.30) is the same as equation (3.6) with $M = T^1_kQ$ and $H = E_L$. Thus from the discussion about existence of solutions of equation (3.6) (see section 3.2), we obtain that in this particular case, the set $\mathfrak{X}^k_L(T^1_kQ)$ is nonempty. ◇

6.4 The Legendre transformation and the equivalence between k-symplectic Hamiltonian and Lagrangian formulations

In this section we shall describe the connection between the Hamiltonian and Lagrangian formulations of classical field theories in the k-symplectic setting.

Definition 6.12. Let $L \in \mathcal{C}^\infty(T^1_kQ)$ be a Lagrangian. The ***Legendre transformation*** for L is the map $FL : T^1_kQ \to (T^1_k)^*Q$ defined as follows:

$$FL(\mathrm{v}_q) = ([FL(\mathrm{v}_q)]^1, \dots, [FL(\mathrm{v}_q)]^k)$$

where

$$[FL(\mathrm{v}_q)]^\alpha(u_q) = \frac{d}{ds}\Big|_{s=0} L\left(v_{1q}, \dots, v_{\alpha q} + su_q, \dots, v_{kq}\right) ,$$

for $1 \le \alpha \le k$ and $u_q \in T_qQ$, $\mathrm{v}_q = (v_{1q}, \dots, v_{kq}) \in T^1_kQ$.

Using natural coordinates (q^i, v^i_α) on T^1_kQ and (q^i, p^α_i) on $(T^1_k)^*Q$, the local expression of the Legendre map is

$$\begin{array}{rccl} FL\colon & T^1_kQ & \to & (T^1_k)^*Q \\ & (q^i, v^i_\alpha) & \longrightarrow & \left(q^i, \dfrac{\partial L}{\partial v^i_\alpha}\right). \end{array} \tag{6.36}$$

The Jacobian matrix of FL is the following matrix of order $n(k+1)$,

$$\begin{pmatrix} I_n & 0 & \cdots & 0 \\ \dfrac{\partial^2 L}{\partial q^i \partial v^j_1} & \dfrac{\partial^2 L}{\partial v^i_1 \partial v^j_1} & \cdots & \dfrac{\partial^2 L}{\partial v^i_k \partial v^j_1} \\ \vdots & \vdots & & \vdots \\ \dfrac{\partial^2 L}{\partial q^i \partial v^j_k} & \dfrac{\partial^2 L}{\partial v^i_1 \partial v^j_k} & \cdots & \dfrac{\partial^2 L}{\partial v^i_k \partial v^j_k} \end{pmatrix}$$

where I_n is the identity matrix of order n and $1 \le i, j \le n$. Thus we deduce that FL is a local diffeomorphism if and only if

$$\det \begin{pmatrix} \dfrac{\partial^2 L}{\partial v^i_1 \partial v^j_1} & \cdots & \dfrac{\partial^2 L}{\partial v^i_k \partial v^j_1} \\ \vdots & & \vdots \\ \dfrac{\partial^2 L}{\partial v^i_1 \partial v^j_k} & \cdots & \dfrac{\partial^2 L}{\partial v^i_k \partial v^j_k} \end{pmatrix} \neq 0$$

with $1 \leq i, j \leq n$.

Definition 6.13. A Lagrangian function $L : T^1_k Q \longrightarrow \mathbb{R}$ is said to be ***regular*** (resp. ***hyperregular***) if the Legendre map FL is a local diffeomorphism (resp. global). In other case L is said to be ***singular***.

The Poincaré-Cartan forms $\theta^\alpha_L, \omega^\alpha_L$, with $1 \leq \alpha \leq k$ (defined in section 6.3.1) are related with the canonical forms $\theta^\alpha, \omega^\alpha$ of $(T^1_k)^*Q$ (defined in section 2.1), using the Legendre map FL.

Lemma 6.2. *For all $1 \leq \alpha \leq k$ one obtains*

$$\theta^\alpha_L = FL^*\theta^\alpha \,, \quad \omega^\alpha_L = FL^*\omega^\alpha \,. \tag{6.37}$$

Proof. It is a direct consequence of the local expressions (2.6), (6.23) and (6.24) of θ^α, ω^α and ω^α_L and the local expression of the Legendre map (6.36). □

Consider $V = \ker(\tau^k)_*$ the vertical distribution of the bundle $\tau^k\colon T^1_k Q \to Q$, then we obtain the following characterization of a regular Lagrangian (the proof of this result can be found in [Merino (1997)])

Prop 6.5. Let $L \in \mathcal{C}^\infty(T^1_k Q)$ be a Lagrangian function. L is regular if and only if $(\omega^1_L, \dots, \omega^k_L, V)$ is a k-symplectic structure on $T^1_k Q$.

Therefore one can state the following theorem:

Theorem 6.2. *Given a Lagrangian function $L\colon T^1_k Q \to \mathbb{R}$, the following conditions are equivalent:*

(1) *L is regular.*

(2) $\det \left(\dfrac{\partial^2 L}{\partial v^i_\alpha \partial v^j_\beta} \right) \neq 0$ *with $1 \leq i, j \leq n$ and $1 \leq \alpha, \beta \leq k$.*

(3) *FL is a local k-symplectomorphism.*

Now we restrict ourselves to the case of hyperregular Lagrangians. In this case the Legendre map FL is a global diffeomorphism and thus we can define a Hamiltonian function $H : (T^1_k)^*Q \to \mathbb{R}$ by

$$H = (FL^{-1})^* E_L = E_L \circ FL^{-1}$$

where FL^{-1} is the inverse diffeomorphism of FL.

In these conditions, we can state the equivalence between both Hamiltonian and Lagrangian formalisms.

Theorem 6.3. *Let $L\colon T^1_kQ \to \mathbb{R}$ be a hyperregular Lagrangian then:*

(1) $\mathbf{X} = (X_1,\dots,X_k) \in \mathfrak{X}^k_L(T^1_kQ)$ *if and only if* $(T^1_kFL)(\mathbf{X}) = (FL_*X_1,\dots, FL_*X_k) \in \mathfrak{X}^k_H((T^1_k)^*Q)$ *where* $H = E_L \circ FL^{-1}$.

(2) *There exists a bijective correspondence between the set of maps* $\phi\colon \mathbb{R}^k \to Q$ *such that* $\phi^{(1)}$ *is an integral section of some* $(X_1,\dots,X_k) \in \mathfrak{X}^k_L(T^1_kQ)$ *and the set of maps* $\psi\colon \mathbb{R}^k \to (T^1_k)^*Q$, *which are integral section of some* $(Y_1,\dots,Y_k) \in \mathfrak{X}^k_H((T^1_k)^*Q)$, *being* $H = (FL^{-1})^*E_L$.

Proof. **(1)** Given FL therefore we can consider the canonical prolongation T^1_kFL following the definition of section 6.1.2. Thus given a k-vector field $\mathbf{X} = (X_1,\dots,X_k) \in \mathfrak{X}^k_L(T^1_kQ)$, one can define a k-vector field on $(T^1_k)^*Q$ using the following diagram

$$\begin{array}{ccc} T^1_kQ & \xrightarrow{FL} & (T^1_k)^*Q \\ \mathbf{X}\Big\downarrow & & \Big\downarrow (T^1_kFL)(\mathbf{X}) \\ T^1_k(T^1_kQ) & \xrightarrow{T^1_kFL} & T^1_k((T^1_k)^*Q) \end{array}$$

that is, for each $1 \le \alpha \le k$, we consider the vector field on $(T^1_k)^*Q$, $FL_*(X_\alpha)$.

We now consider the function $H = E_L \circ FL^{-1} = (FL^{-1})^*E_L$, then

$$(T^1_kFL)(\mathbf{X}) = (FL_*(X_1),\dots,FL_*(X_k)) \in \mathfrak{X}^k_H((T^1_k)^*Q)$$

if and only if

$$\sum_{\alpha=1}^k \iota_{FL_*(X_\alpha)}\omega^\alpha - d\Big((FL^{-1})^*E_L\Big) = 0\,.$$

Since FL is a diffeomorphism, this is equivalent to

$$0 = FL^*\Big(\sum_{\alpha=1}^k \iota_{FL_*X_\alpha}\omega^\alpha - d(FL^{-1})^*E_L\Big) = \sum_{\alpha=1}^k \iota_{X_\alpha}(FL)^*\omega^\alpha - dE_L$$

and from (6.37), this fact occurs if and only if $\mathbf{X} \in \mathfrak{X}^k_L(T^1_kQ)$.

Finally, observe that since FL is a diffeomorphism, T^1_kFL is also a diffeomorphism, and then any k-vector field on $(T^1_k)^*Q$ is of the type $T^1_kFL(\mathbf{X})$ for some $\mathbf{X} \in \mathfrak{X}^k(T^1_kQ)$.

(2) Let $\phi\colon \mathbb{R}^k \to Q$ be a map such that its first prolongation $\phi^{(1)}$ is an integral section of some $\mathbf{X} = (X_1,\dots,X_k) \in \mathfrak{X}^k_L(T^1_kQ)$, then the map $\psi = FL \circ \phi^{(1)}$ is an integral section of $T^1_kFL(\mathbf{X}) = (FL_*(X_1),\dots,FL_*(X_k))$. Since we have proved that $T^1_kFL(\mathbf{X}) \in \mathfrak{X}^k_H((T^1_k)^*Q)$, we obtain the first part of item 2.

The converse is similar, if we consider that any k-vector field on $(T^1_k)^*Q$ is of the type $T^1_k FL(\mathbf{X})$ for some $\mathbf{X} \in \mathfrak{X}^k(T^1_k Q)$. Thus given $\psi\colon \mathbb{R}^k \to (T^1_k)^*Q$ integral section of any $(Y_1, \dots, Y_k) \in \mathfrak{X}^k_H((T^1_k)^*Q)$, there exists a k-vector field $\mathbf{X} \in \mathfrak{X}^k_L(T^1_k Q)$ such that $T^1_k FL(\mathbf{X}) = (Y_1, \dots, Y_k)$. Finally, the map ψ corresponds with $\phi^{(1)}$ where $\phi = \pi^k \circ \psi$.

$\square$

Chapter 7

Examples

In this chapter we describe several physical examples using the k-symplectic formulation developed in this part of the book. In [Muñoz-Lecanda, Salgado and Vilariño (2009)] one can find several of these examples. Previously, we recall the geometric version of the Hamiltonian and Lagrangian approaches for classical field theories and its correspondence with the case $k = 1$.

	k-symplectic formalism	Symplectic formalism ($k = 1$) (classical mechanics)
Hamiltonian formalism	$\sum_{\alpha=1}^{k} i_{X_\alpha}\omega^\alpha = dH$ $\mathbf{X} \in \mathfrak{X}^k(M)$ M k-symplectic manifold	$i_X\omega = dH$ $X \in \mathfrak{X}(M)$ M symplectic manifold
Lagrangian formalism	$\sum_{\alpha=1}^{k} i_{X_\alpha}\omega_L^\alpha = dE_L$ $\mathbf{X} \in \mathfrak{X}^k(T_k^1Q)$	$i_X\omega_L = dE_L$ $X \in \mathfrak{X}(TQ)$

As before, the canonical coordinates in $\mathbb{R}^k$ are denoted by $(x^1, \ldots, x^k)$. Moreover, we shall use the following notation for the partial derivatives of a map $\phi\colon \mathbb{R}^k \to Q$:

$$\partial_\alpha \phi^i = \frac{\partial \phi^i}{\partial x^\alpha}, \quad \partial_{\alpha\beta}\phi^i = \frac{\partial^2 \phi^i}{\partial x^\alpha \partial x^\beta}, \tag{7.1}$$

where $1 \leq \alpha, \beta \leq k$ and $1 \leq i \leq n$.

7.1 Electrostatic equations

We now consider the study of electrostatic in a 3-dimensional manifold M with coordinates (x^1, x^2, x^3), for instance $M = \mathbb{R}^3$. We assume that M is a Riemannian manifold with a metric g with components $g_{\alpha\beta}(x)$ where $1 \leq \alpha, \beta \leq 3$.

The equations of electrostatics are (see [Durand (1964); Kijowski and Tulczyjew (1979)]):

$$\begin{aligned} E &= \star d\psi, \\ dE &= -4\pi\rho, \end{aligned} \tag{7.2}$$

where $\star$ is the Hodge operator[1] associated with the metric g, ψ is a scalar field $\psi\colon \mathbb{R}^3 \to \mathbb{R}$ given the electric potential on $\mathbb{R}^3$ and $E = (\psi^1, \psi^2, \psi^3)\colon \mathbb{R}^3 \to \mathbb{R}^3$ is a vector field which gives the electric field on $\mathbb{R}^3$ and such that it can be interpreted by the 2-form on $\mathbb{R}^3$ given by

$$E = \psi^1 dx^2 \wedge dx^3 + \psi^2 dx^3 \wedge dx^1 + \psi^3 dx^1 \wedge dx^2 ,$$

and ρ is the 3-form on $\mathbb{R}^3$ representing a fixed charge density

$$\rho(x) = \sqrt{g} r(x) dx^1 \wedge dx^2 \wedge dx^3 , \tag{7.3}$$

where $g = |\det g_{\alpha\beta}|$.

In terms of local coordinates, the above system of equations (7.2) reads:

$$\begin{aligned} \psi^\alpha &= \sqrt{g} g^{\alpha\beta} \frac{\partial \psi}{\partial x^\beta} , \\ \sum_{\alpha=1}^{k} \frac{\partial \psi^\alpha}{\partial x^\alpha} &= -4\pi \sqrt{g} r , \end{aligned} \tag{7.4}$$

where r is the scalar function defined by the equation $r = \star\rho$, or equivalently, by (7.3).

Suppose that g is the Euclidean metric on $\mathbb{R}^3$, thus the above equations can be written as follows:

$$\begin{aligned} \psi^\alpha &= \frac{\partial \psi}{\partial x^\alpha}, \\ -\left(\frac{\partial \psi^1}{\partial x^1} + \frac{\partial \psi^2}{\partial x^2} + \frac{\partial \psi^3}{\partial x^3}\right) &= 4\pi r . \end{aligned} \tag{7.5}$$

[1] In general, on an orientable n-manifold with a Riemannian metric g, the *Hodge operator* $\star\colon \Omega^k(M) \to \Omega^{n-k}(M)$ is a linear operator that for every $\nu, \eta \in \Omega^k(M)$

$$\nu \wedge \star\eta = g(\nu, \eta) \mathrm{d}vol_g ,$$

where $\mathrm{d}vol_g$ is the Riemann volume. In local coordinates we have

$$\nu = \nu_{i_1 \ldots i_k} \mathrm{d}x^{i_1} \wedge \mathrm{d}x^{i_k} , \eta = \eta_{j_1 \ldots j_k} \mathrm{d}x^{j_1} \wedge \mathrm{d}x^{j_k} ,\ g(\nu, \eta) = \nu_{i_1 \ldots i_k} \eta_{j_1 \ldots j_k} g^{i_1 j_1} \ldots g^{i_k j_k} ,$$

and $\mathrm{d}vol_g = \sqrt{|det(g_{ij})|} dx^1 \wedge \ldots \wedge dx^n$, (g^{ij}) being the inverse of the metric matrix (g_{ij}). For more details see, for instance, [Du, Hao, Hu, Hui, Shi, Wang and Wu (2011)].

As we have seen in section 3.3, these equations can be obtained from the 3-symplectic equation

$$\iota_{X_1}\omega^1 + \iota_{X_2}\omega^2 + \iota_{X_3}\omega^3 = dH$$

with $H\colon (T^1_3)^*\mathbb{R} \to \mathbb{R}$ the Hamiltonian defined in (3.15).

7.2 Wave equation

Consider the $(n+1)$-symplectic Hamiltonian equation

$$\sum_{\alpha=1}^{n+1} \iota_{X_\alpha}\omega^\alpha = \mathrm{d}H\,, \tag{7.6}$$

associated to the Hamiltonian function

$$\begin{aligned} H\colon \quad (T^1_{n+1})^*\mathbb{R} \quad &\to \mathbb{R} \\ (q,p^1,\dots,p^{n+1}) &\mapsto \frac{1}{2}\Big((p^{n+1})^2 - \frac{1}{c^2}\sum_{\alpha=1}^{n}(p^\alpha)^2\Big)\,, \end{aligned} \tag{7.7}$$

where $(q,p^1,\dots,p^{n+1})$ are the canonical coordinates on $(T^1_{n+1})^*\mathbb{R}$ introduced in section 2.1.

Let $\mathbf{X} = (X_1,\dots,X_{n+1})$ be an integrable $(n+1)$-vector field which is a solution of equation (7.6); then since

$$\frac{\partial H}{\partial q} = 0, \quad \frac{\partial H}{\partial p^\alpha} = -\frac{1}{c^2}p^\alpha,\ 1 \le \alpha \le n \quad \text{and} \quad \frac{\partial H}{\partial p^{n+1}} = p^{n+1}$$

we deduce, from (3.7), that each X_α is locally given by

$$\begin{aligned} X_\alpha \quad &= -\frac{1}{c^2}p^\alpha\frac{\partial}{\partial q} + (X_\alpha)^\beta\frac{\partial}{\partial p^\beta}, \quad 1 \le \alpha \le n, \\ X_{n+1} &= p^{n+1}\frac{\partial}{\partial q} + (X_{n+1})^\beta\frac{\partial}{\partial p^\beta}\,, \end{aligned} \tag{7.8}$$

and the components $(X_\alpha)^\beta$ satisfy $\displaystyle\sum_{\alpha=1}^{n+1}(X_\alpha)^\alpha = 0$.

Remark 7.1. In this particular case the integrability condition of $\mathbf{X}$ is equivalent to the following local conditions:

$$\begin{aligned} (X_\alpha)^\beta &= (X_\beta)^\alpha, \\ (X_\alpha)^{n+1} &= -{}^1\!/_{c^2}(X_{n+1})^\alpha, \\ X_\alpha\Big((X_\beta)^\gamma\Big) &= X_\beta\Big((X_\alpha)^\gamma\Big), \\ X_\alpha\Big((X_{n+1})^\gamma\Big) &= X_{n+1}\Big((X_\alpha)^\gamma\Big), \end{aligned}$$

where $1 \le \alpha,\beta \le n$ and $1 \le \gamma \le n+1$. ◇

We now consider an integral section

$$(x^1,\dots,x^n,t)\to(\psi(x^1,\dots,x^n,t),\psi^1(x^1,\dots,x^n,t),\dots,\psi^{n+1}(x^1,\dots,x^n,t))$$

of the $(n+1)$-vector field $\mathbf{X}=(X_1,\dots,X_{n+1})\in\mathfrak{X}^{n+1}_H(\mathbb{R})$. From (7.8) one deduce that the integral section satisfies

$$\psi^\alpha=-c^2\frac{\partial\psi}{\partial x^\alpha},\quad 1\leq\alpha\leq n, \tag{7.9}$$

$$\psi^{n+1}=\frac{\partial\psi}{\partial t}, \tag{7.10}$$

$$0=\sum_{\alpha=1}^{n}\frac{\partial\psi^\alpha}{\partial x^\alpha}+\frac{\partial\psi^{n+1}}{\partial t}. \tag{7.11}$$

Finally, if we consider the identities (7.9) and (7.10) in (7.11) one deduces that ψ is a solution of

$$\frac{\partial^2\psi}{\partial t^2}=c^2\nabla^2\psi, \tag{7.12}$$

where ∇^2 is the (spatial) Laplacian, i.e., ψ is a solution of the n-dimensional wave equation. Let us recall that a solution of this equation is a scalar function $\psi=\psi(x^1,\dots,x^n,t)$ whose values model the height of a wave at the position $(x^1,\dots,x^n)$ and at the time t.

The Lagrangian counterpart of this example is the following. Consider the Lagrangian $(n+1)$-symplectic equation

$$\sum_{\alpha=1}^{n+1}\iota_{X_\alpha}\omega_L^\alpha=\mathrm{d}E_L, \tag{7.13}$$

associated to the Lagrangian function

$$\begin{aligned}L\colon\quad (T^1_{n+1})^*\mathbb{R}&\to\mathbb{R}\\ (q,v_1,\dots,v_{n+1})&\mapsto\frac{1}{2}\Big((v_{n+1})^2-c^2\sum_{\alpha=1}^{n}v_\alpha^2\Big).\end{aligned} \tag{7.14}$$

where $(q,v_1,\dots,v_{n+1})$ are the canonical coordinates on $T^1_{n+1}\mathbb{R}$.

Let $\mathbf{X}=(X_1,\dots,X_{n+1})$ be an integrable $(n+1)$-vector field solution of equation (7.13), then

$$X_\alpha=v_\alpha\frac{\partial}{\partial q^i}+(X_\alpha)_\beta\frac{\partial}{\partial v_\beta} \tag{7.15}$$

and the components $(X_\alpha)_\beta$ satisfy equations (6.34), which in this case are

$$\begin{aligned} 0 &= \sum_{\alpha,\beta=1}^{n+1} \frac{\partial^2 L}{\partial v_\alpha \partial v_\beta}(X_\alpha)_\beta \\ &= \sum_{\alpha,\beta=1}^{n} \frac{\partial^2 L}{\partial v_\alpha \partial v_\beta}(X_\alpha)_\beta + \frac{\partial^2 L}{\partial v_{n+1}\partial v_{n+1}}(X_{n+1})_{n+1} \qquad (7.16) \\ &= -c^2 \sum_{\alpha=1}^{n}(X_\alpha)_\alpha + (X_{n+1})_{n+1} \end{aligned}$$

since

$$\frac{\partial^2 L}{\partial v_\alpha \partial v_\beta} = -c^2\delta^{\alpha\beta}, \quad 1 \leq \alpha, \beta \leq n, \quad \frac{\partial^2 L}{\partial v_\alpha \partial v_{n+1}} = 0, \quad \frac{\partial^2 L}{\partial v_{n+1}\partial v_{n+1}} = 1\,.$$

Now, if

$$\begin{aligned} \phi^{(1)} : \mathbb{R}^{n+1} &\longrightarrow T^1_{n+1}\mathbb{R} \\ x &\to \phi(x) = \left(\phi(x), \frac{\partial \phi}{\partial x^\alpha}(x)\right) \end{aligned}$$

is an integral section of $\mathbf{X}$, then we deduce from (6.29) and (7.16) that $\phi : \mathbb{R}^{n+1} \to \mathbb{R}$ is a solution of equations (7.12).

7.3 Laplace's equations

On the n-symplectic manifold $(T^1_n)^*\mathbb{R}$ we define the Hamiltonian function

$$\begin{aligned} H : \quad (T^1_n)^*\mathbb{R} &\to \mathbb{R} \\ (q, p^1, \dots, p^n) &\mapsto \frac{1}{2}\left((p^1)^2 + \dots + (p^n)^2\right), \end{aligned}$$

where $(q, p^1, \dots, p^n)$ are canonical coordinates on $(T^1_n)^*\mathbb{R}$. Then

$$\frac{\partial H}{\partial q} = 0, \quad \frac{\partial H}{\partial p^\alpha} = p^\alpha, \qquad (7.17)$$

with $1 \leq \alpha \leq n$.

The n-symplectic Hamiltonian equation (3.6) associated with H is

$$\iota_{X_1}\omega^1 + \dots + \iota_{X_n}\omega^n = dH\,. \qquad (7.18)$$

From (3.7) and (7.17) we deduce that an integrable n-vector field solution of (7.18), has the following local expression:

$$X_\alpha = p^\alpha \frac{\partial}{\partial q} + (X_\alpha)^\beta \frac{\partial}{\partial p^\beta}, \qquad (7.19)$$

and its components satisfy the following equations

$$0 = \sum_{\alpha=1}^{n} (X_\alpha)^\alpha \,, \tag{7.20}$$

$$(X_\alpha)^\beta = (X_\beta)^\alpha, \tag{7.21}$$

$$X_\alpha\Big((X_\beta)^\gamma\Big) = X_\beta\Big((X_\alpha)^\gamma\Big)\,, \tag{7.22}$$

with $1 \leq \alpha, \beta, \gamma \leq n$. Let us observe that (7.21) and (7.22) are the integrability condition of the n-vector field $\mathbf{X} = (X_1, \dots, X_n)$.

If

$$\begin{aligned} \varphi : \mathbb{R}^3 &\longrightarrow (T^1_3)^*\mathbb{R} \\ x &\to \varphi(x) = (\psi(x), \psi^1(x), \psi^2(x), \psi^3(x)) \end{aligned}$$

is an integral section of $(X_1, \dots, X_n)$, then from (7.19) and (7.20) we obtain that

$$\psi^\alpha = \frac{\partial \psi}{\partial x^\alpha}\,,$$

$$\sum_{\alpha=1}^{n} \frac{\partial \psi^\alpha}{\partial x^\alpha} = 0\,.$$

Therefore, ψ is a solution of

$$\frac{\partial^2 \psi}{\partial (x^1)^2} + \ldots + \frac{\partial^2 \psi}{\partial (x^n)^2} = 0\,, \tag{7.23}$$

that is, ψ is a solution of Laplace's equations [Olver (1986, 2007)].

Let $(X_1, \dots, X_n)$ be an n-vector field on $T^1_n\mathbb{R}$, with coordinates $(q, v_1, \dots, v_n)$, which is a solution of

$$\iota_{X_1}\omega_L^1 + \ldots + \iota_{X_n}\omega_L^n = dE_L\,, \tag{7.24}$$

where L is the regular Lagrangian

$$\begin{aligned} L: \quad T^1_n\mathbb{R} &\to \mathbb{R} \\ (q, v_1, \dots, v_n) &\mapsto \frac{1}{2}((v_1)^2 + \ldots + (v_n)^2)\,. \end{aligned}$$

From (6.1), and taking into account that

$$\frac{\partial L}{\partial q} = 0\,, \quad \frac{\partial L}{\partial v_\alpha} = v_\alpha\,,$$

with $1 \leq \alpha \leq k$, we obtain that if ϕ is a solution of the n-vector field $(X_1, \dots, X_n)$ on $T^1_n\mathbb{R}$, then ϕ satisfies

$$\partial_{11}\phi + \ldots + \partial_{nn}\phi = 0\,,$$

or equivalently

$$\nabla^2 \phi = 0\,,$$

which is the *Laplace equation* (7.23). Thus equations (7.24) can be considered as the geometric version of Laplace's equations.

Remark 7.2. The solutions of the Laplace equations are important in many fields of science, for instance, electromagnetism, astronomy and fluid dynamics, because they describe the behavior of electric, gravitational and fluid potentials. The solutions of Laplace's equations are called harmonic functions. ◇

7.4 Sine-Gordon equation

Define the Hamiltonian function

$$\begin{aligned} H\colon\ (T^1_2)^*\mathbb{R} &\to \mathbb{R} \\ (q,p^1,p^2) &\mapsto \frac{1}{2}\left((p^1)^2 - \frac{1}{a^2}(p^2)^2\right) - \Omega^2 \cos q \end{aligned},$$

a^2 and Ω^2 being two positive constants.

Consider the 2-symplectic Hamiltonian equations associated to this Hamiltonian, i.e.,

$$\iota_{X_1}\omega^1 + \iota_{X_2}\omega^2 = dH\,, \tag{7.25}$$

and let $\mathbf{X} = (X_1, X_2)$ be a solution.

In canonical coordinates (q, p^1, p^2) on $(T^1_2)^*Q$, a solution $\mathbf{X}$ has the following local expression

$$\begin{aligned} X_1 &= p^1\frac{\partial}{\partial q} + (X_1)^1\frac{\partial}{\partial p^1} + (X_1)^2\frac{\partial}{\partial p^2}, \\ X_2 &= -\frac{1}{a^2}p^2\frac{\partial}{\partial q} + (X_2)^1\frac{\partial}{\partial p^1} + (X_2)^2\frac{\partial}{\partial p^2}, \end{aligned} \tag{7.26}$$

where the functions $(X_\alpha)^\beta$ satisfy $(X_1)^1 + (X_2)^2 = -\Omega^2 \sin q$.

If (X_1, X_2) is an integrable 2-vector field, that is $[X_1, X_2] = 0$, then the functions $(X_1)^2$ and $(X_2)^1$ satisfy $(X_2)^1 = -{}^1\!/_{a^2}(X_1)^2$.

Let $\varphi\colon \mathbb{R}^2 \to (T^1_2)^*\mathbb{R}$, $\varphi(x) = (\psi(x), \psi^1(x), \psi^2(x))$ be an integral section of the 2-vector field $\mathbf{X}$. Then from (7.26) one has that φ satisfies

$$\psi^1 = \frac{\partial \psi}{\partial x^1}\,, \psi^2 = -a^2\frac{\partial \psi}{\partial x^2}\,, \frac{\partial \psi^1}{\partial x^1} + \frac{\partial \psi^2}{\partial x^2} = -\Omega^2 \sin\psi\,, \tag{7.27}$$

and hence $\psi\colon \mathbb{R}^2 \to \mathbb{R}$ is a solution of

$$\frac{\partial^2 \psi}{\partial (x^1)^2} - a^2 \frac{\partial^2 \psi}{\partial (x^2)^2} + \Omega^2 \sin \psi = 0\,, \tag{7.28}$$

that is, ψ is a solution of the *Sine-Gordon equation* (see [José and Saletan (1998)]).

Remark 7.3. The Sine-Gordon equation was known in the 19th century, but the equation grew greatly in importance when it was realized that it led to solutions "kink" and "antikink" with the collisional properties of solitons [Perring and Skyrme (1962)]. This equation also appears in other physical applications [Barone, Esposito, Magee and Scott (1971); Bishop and Schneider (1978); Davydov (1985); Gibbon, James and Moroz (1979); Infeld and Rowlands (2000)], including the motion of rigid pendula attached to a stretched wire, and dislocations in crystals. ◇

This equation (7.28) can also be obtained from the Lagrangian approach if we consider the 2-symplectic equation

$$\imath_{X_1}\omega_L^1 + \imath_{X_2}\omega_L^2 = dE_L\,, \tag{7.29}$$

where (X_1, X_2) is a 2-vector field on $T_2^1\mathbb{R}$ and the Lagrangian is the function

$$L(q, v_1, v_2) = \frac{1}{2}((v_1)^2 - a^2(v_2)^2) - \Omega^2(1 - \cos(q))$$

a^2 and Ω^2 being two positive constants.

Thus we have

$$\frac{\partial L}{\partial q} = -\Omega^2 \sin(q)\,, \quad \frac{\partial L}{\partial v_1} = v_1\,, \quad \frac{\partial L}{\partial v_2} = -a^2 v_2\,. \tag{7.30}$$

From (6.1) and (7.30), we know that if ϕ is a solution of (X_1, X_2) then

$$0 = \partial_{11}\phi - a^2\partial_{22}\phi + \Omega^2 \sin\phi\,,$$

that is, ϕ is a solution of the Sine-Gordon equation (7.28).

7.5 Ginzburg-Landau's equation

Let us consider the Hamiltonian function

$$\begin{aligned} H\colon\ & (T_2^1)^*\mathbb{R} \to \mathbb{R} \\ & (q, p^1, p^2) \mapsto \frac{1}{2}\Big((p^1)^2 - \frac{1}{a^2}(p^2)^2\Big) - \lambda(q^2 - 1)^2 \end{aligned}$$

where a and λ are supposed to denote constant quantities. Then

$$\frac{\partial H}{\partial q} = -4\lambda q(q^2-1), \quad \frac{\partial H}{\partial p^1} = p^1, \quad \frac{\partial H}{\partial p^2} = -\frac{1}{a^2}p^2 .$$

Consider the 2-symplectic Hamiltonian equations associated to this Hamiltonian, i.e.,

$$\iota_{X_1}\omega^1 + \iota_{X_2}\omega^2 = dH\,, \tag{7.31}$$

and let $\mathbf{X} = (X_1, X_2)$ be a solution.

In the canonical coordinates (q, p^1, p^2) on $(T^1_2)^*\mathbb{R}$, a 2-vector field $\mathbf{X}$ solution of (7.31) has the following local expression

$$\begin{aligned} X_1 &= p^1\frac{\partial}{\partial q} + (X_1)^1\frac{\partial}{\partial p^1} + (X_1)^2\frac{\partial}{\partial p^2}, \\ X_2 &= -\frac{1}{a^2}p^2\frac{\partial}{\partial q} + (X_2)^1\frac{\partial}{\partial p^1} + (X_2)^2\frac{\partial}{\partial p^2}, \end{aligned} \tag{7.32}$$

where the functions $(X_\alpha)^\beta$ satisfy $(X_1)^1 + (X_2^2) = 4\lambda q(q^2-1)$.

A necessary condition for the integrability of the 2-vector field (X_1, X_2) is that $(X_2)^1 = -{}^1\!/\!_{a^2}(X_1)^2$.

Let $\varphi\colon \mathbb{R}^2 \to (T^1_2)^*\mathbb{R}$ be an integral section of the 2-vector field $\mathbf{X}$ with components $\varphi(x) = (\psi(x),\ \psi^1(x), \psi^2(x))$. Then from (7.32) one obtains that φ satisfies

$$\psi^1 = \frac{\partial\psi}{\partial x^1}\,, \quad \psi^2 = -a^2\frac{\partial\psi}{\partial x^2}\,, \quad \frac{\partial\psi^1}{\partial x^1} + \frac{\partial\psi^2}{\partial x^2} = 4\lambda\psi(\psi^2-1)\,. \tag{7.33}$$

Hence ψ is a solution of

$$\frac{\partial^2\psi}{\partial(x^1)^2} - a^2\frac{\partial^2\psi}{\partial(x^2)^2} - 4\lambda\psi(\psi^2-1) = 0\,, \tag{7.34}$$

that is, ψ is a solution of *Ginzburg-Landau's equation.*

Next, let us consider the Lagrangian

$$\begin{aligned} L\colon\ T^1_2\mathbb{R} &\equiv T\mathbb{R}\oplus T\mathbb{R} \to \mathbb{R} \\ (q, v_1, v_2) &\mapsto \frac{1}{2}[(v_1)^2 - a^2(v_2)^2] + \lambda(q^2-1)^2\,. \end{aligned}$$

Here a and λ are supposed to denote *constant* quantities. Then

$$\frac{\partial L}{\partial q} = 4\lambda q(q^2-1)\,, \quad \frac{\partial L}{\partial v_1} = v_1\,, \quad \frac{\partial L}{\partial v_2} = -a^2 v_2\,. \tag{7.35}$$

Let (X_1, X_2) be a 2-vector field on $T^1_2\mathbb{R}$ solution of

$$\iota_{X_1}\omega^1_L + \iota_{X_2}\omega^2_L = dE_L\,.$$

If ϕ is a solution of (X_1, X_2), then from (6.1) and (7.35) we obtain that ϕ satisfies the equation

$$0 = \partial_{11}\phi - a^2\partial_{22}\phi - 4\lambda\phi(\phi^2 - 1)\,,$$

which is the Ginzburg-Landau equation (7.34).

Remark 7.4. The phenomenological Ginzburg-Landau theory (1950) is a mathematical theory used for modeling superconductivity [Ginzburg and Landau (1950)]. ◇

7.6 k-symplectic quadratic systems

Many Hamiltonian and Lagrangian systems in field theories are of "quadratic" type and they can be modeled as follows.

Consider the canonical model of k-symplectic manifold $((T^1_k)^*Q, \omega^\alpha, V)$. Let $g_1, \dots, g_k$ be k semi-Riemannian metrics in Q. For each $q \in Q$ and for each $1 \le \alpha \le k$ we have the following linear isomorphisms:

$$\begin{array}{rl} g^\flat_\alpha \colon & T_qQ \to T^*_qQ \\ & v \;\;\mapsto \iota_v g_\alpha \end{array},$$

and then we introduce the dual metric g^*_α of g_α, defined as follows

$$g^*_\alpha(\nu_q, \gamma_q) \colon = g_\alpha\big((g^\flat_\alpha)^{-1}(\nu_q), (g^\flat_\alpha)^{-1}(\gamma_q)\big)\,,$$

for each $\nu_q, \gamma_q \in T^*_qQ$ and $1 \le \alpha \le k$.

We can define a function $K \in \mathcal{C}^\infty((T^1_k)^*Q)$ as follows: for every $(\nu_{1_q}, \dots, \nu_{k_q}) \in (T^1_k)^*Q$,

$$K(\nu_{1_q}, \dots, \nu_{k_q}) = \frac{1}{2}\sum_{\alpha=1}^{k} g^*_\alpha(\nu_{\alpha_q}, \nu_{\alpha_q})\,.$$

Then, if $V \in \mathcal{C}^\infty(Q)$ we define the Hamiltonian function $H \in \mathcal{C}^\infty((T^1_k)^*Q)$ of "quadratic" type as follows

$$H = K + (\pi^k)^*V\,.$$

Using canonical coordinates (q^i, p^α_i) on $(T^1_k)^*Q$, the local expression of H is

$$H(q^i, p^\alpha_i) = \frac{1}{2}\sum_{\alpha=1}^{k} g^{ij}_\alpha(q^m)p^\alpha_i p^\alpha_j + V(q^m)\,,$$

where g_α^{ij} denote the coefficients of the matrix associated to g_α^*. Then

$$dH = \sum_{\alpha=1}^{k} \left[\left(\frac{1}{2} \frac{\partial g_\alpha^{ij}}{\partial q^k} p_i^\alpha p_j^\alpha + \frac{\partial V}{\partial q^k} \right) dq^k + (g_\alpha^{ij} p_i^\alpha) dp_j^\alpha \right] .$$

Consider now the k-symplectic Hamiltonian field equation (3.6) associated with the above Hamiltonian function, i.e.,

$$\sum_{\alpha=1}^{k} \iota_{X_\alpha} \omega^\alpha = dH .$$

If a k-vector field $\mathbf{X} = (X_1, \ldots, X_k)$ is solution of this equation then each X_α has the following local expression (for each α fixed):

$$X_\alpha = g_\alpha^{ij} p_j^\alpha \frac{\partial}{\partial q^i} + (X_\alpha)_i^\beta \frac{\partial}{\partial p_i^\beta} \tag{7.36}$$

and its components $(X_\alpha)_i^\beta$ satisfy

$$\sum_{\beta=1}^{k} (X_\beta)_k^\beta = - \left(\frac{1}{2} \frac{\partial g_\beta^{ij}}{\partial q^k} p_i^\beta p_j^\beta + \frac{\partial V}{\partial q^k} \right) . \tag{7.37}$$

We now assume that $\mathbf{X}$ is integrable and

$$\begin{aligned} \varphi : \mathbb{R}^k &\longrightarrow (T_k^1)^* Q \\ x &\rightarrow \varphi(x) = (\psi^i(x), \psi_i^\alpha(x)) \end{aligned}$$

is an integral section of $\mathbf{X}$ then

$$X_\alpha(\varphi(x)) = \varphi_*(x) \left(\frac{\partial}{\partial x^\alpha}\Big|_x \right) = \frac{\partial \psi^i}{\partial x^\alpha}\Big|_x \frac{\partial}{\partial q^i}\Big|_{\varphi(x)} + \frac{\partial \psi_i^\beta}{\partial x^\alpha}\Big|_x \frac{\partial}{\partial p_i^\beta}\Big|_{\varphi(x)} . \tag{7.38}$$

Thus, from (7.36)–(7.38) we obtain that φ is a solution of the following Hamilton-De Donder-Weyl equations

$$\frac{\partial \psi^i}{\partial x^\alpha} = g_\alpha^{ij} \psi_j^\alpha , \quad (\alpha \text{ fixed})$$

$$\sum_{\beta=1}^{k} \frac{\partial \psi_l^\beta}{\partial x^\beta} = - \left(\frac{1}{2} \frac{\partial g_\beta^{ij}}{\partial q^l} \psi_i^\beta \psi_j^\beta + \frac{\partial V}{\partial q^l} \right) .$$

In the Lagrangian approach we obtain a similar description. In fact, we consider the tangent bundle of k^1-velocities and let $g_1, \ldots, g_k$ be k semi-Riemannian metrics in Q.

We can define a function $K \in \mathcal{C}^\infty(T^1_kQ)$ as follows: for every element $\mathrm{v}_q = (v_{1_q}, \dots, v_{k_q}) \in T^1_kQ$,

$$K(\mathrm{v}_q) = \frac{1}{2}\sum_{\alpha=1}^{k} g_\alpha(v_{\alpha_q}, v_{\alpha_q}) .$$

Then, $V \in \mathcal{C}^\infty(Q)$ defines the k-symplectic Lagrangian function $L \in \mathcal{C}^\infty(T^1_kQ)$ of "quadratic" type as follows

$$L = K - (\tau^k)^* V .$$

Using canonical coordinates (q^i, v^i_α) on T^1_kQ, the local expression of L is

$$L(q^i, v^i_\alpha) = \frac{1}{2}\sum_{\alpha=1}^{k} g^\alpha_{ij}(q^m) v^i_\alpha\, v^j_\alpha - V(q^m) ,$$

where g^α_{ij} denote the coefficients of the matrix associated to g_α.

Consider now the k-symplectic Lagrangian field equation associated with the above Lagrangian function, i.e.,

$$\sum_{\alpha=1}^{k} \iota_{X_\alpha} \omega^\alpha_L = dE_L .$$

If a k-vector field $\mathbf{X} = (X_1, \dots, X_k)$ is a solution of this equation, i.e., if $\mathbf{X} \in \mathfrak{X}^k_L(T^1_kQ)$ then, since L is regular, each X_α has the following local expression (for each α fixed):

$$X_\alpha = v^i_\alpha \frac{\partial}{\partial q^i} + (X_\alpha)^i_\beta \frac{\partial}{\partial v^i_\beta} \tag{7.39}$$

and its components $(X_\alpha)^i_\beta$ satisfy equations (6.34), that in this case are

$$\frac{\partial g^\alpha_{il}}{\partial q^j} v^l_\alpha v^j_\alpha + g^\alpha_{ij}\,(X_\alpha)^j_\alpha = \frac{1}{2}\frac{\partial g^\alpha_{lm}}{\partial q^i} v^l_\alpha v^m_\alpha - \frac{\partial V}{\partial q^i} .$$

Thus, if the components of the metrics g^α_{il} are constant then

$$g^\alpha_{ij}\,(X_\alpha)^j_\alpha = -\frac{\partial V}{\partial q^i} .$$

Now, if

$$\begin{aligned} \phi^{(1)} : \mathbb{R}^k &\longrightarrow T^1_kQ \\ x &\to \phi(x) = (\phi^i(x), \frac{\partial \phi}{\partial x^\alpha}(x)) \end{aligned}$$

is an integral section of $\mathbf{X} \in \mathfrak{X}^k_L(T^1_kQ)$ then $\phi : \mathbb{R}^k \to Q$ is a solution the following Euler-Lagrange equations

$$g^{\alpha}_{ij} \frac{\partial^2 \phi^j}{\partial x^\alpha \partial x^\beta} = -\frac{\partial V}{\partial q^i} .$$

Remark 7.5. The examples of the previous subsections can be considered a particular case of this situation.

- ***The electrostatic equations*** correspond with the case $Q = \mathbb{R}\,(n = 1)$, $k = 3$, the function $V \in \mathcal{C}^\infty(\mathbb{R})$ is $V(q) = 4\pi r$ and the semi-Riemannian metrics in $\mathbb{R}$,

  $$g_\alpha = dq^2 \quad 1 \le \alpha \le 3 ,$$

  q being the standard coordinate in $\mathbb{R}$.
- ***The wave equation*** corresponds to the case $Q = \mathbb{R}\,(n = 1)$, $k = n+1$, the function $V = 0$ and the semi-Riemannian metrics in $\mathbb{R}$,

  $$g_\alpha = -c^2 dq^2, \quad 1 \le \alpha \le n \text{ and } g_{n+1} = dq^2 ,$$

  q being the standard coordinate in $\mathbb{R}$.
- ***Laplace's equations*** corresponds with the case $Q = \mathbb{R}, k = n, V(q) = 0$ and the semi-Riemaniann metrics $g_\alpha = dq^2$.
- ***The Sine-Gordon equation*** corresponds with the case $Q = \mathbb{R}$, $k = 2$, $V(q) = -\Omega^2 \cos q$, and the semi-Riemannian metrics in $\mathbb{R}$,

  $$g_1 = dq^2 \text{ and } g_2 = -a^2 dq^2 ,$$

  q being the standard coordinate in $\mathbb{R}$.
- In the case of ***Ginzburg-Landau's equation***, $Q = \mathbb{R}$, $k = 2$, $V(q) = -\lambda(q^2 - 1)^2$ and the semi-Riemannian metrics in $\mathbb{R}$,

  $$g_1 = dq^2 \text{ and } g_2 = -a^2 dq^2 ,$$

  q being the standard coordinate in $\mathbb{R}$.

7.7 Navier's equations

We consider the equation (7.29) but with $Q = \mathbb{R}^2$ and Lagrangian $L\colon T\mathbb{R}^2 \oplus T\mathbb{R}^2 \to \mathbb{R}$ given by

$$L(q^1, q^2, v^1_1, v^1_2, v^2_1, v^2_2) = (\frac{1}{2}\lambda + \mu)[(v^1_1)^2 + (v^2_2)^2] + \frac{1}{2}\mu[(v^1_2)^2 + (v^2_1)^2] + (\lambda + \mu) v^1_1 v^2_2 .$$

This Lagrangian is regular if $\mu \neq 0$ and $\lambda \neq -(3/2)\mu$. In this case we obtain:

$$\begin{aligned}
&\frac{\partial L}{\partial q} = 0\,, \\
&\frac{\partial L}{\partial v_1^1} = (\lambda + 2\mu)v_1^1 + (\lambda+\mu)v_2^2\,, \quad \frac{\partial L}{\partial v_2^1} = \mu v_2^1 \\
&\frac{\partial L}{\partial v_1^2} = \mu v_1^2\,, \quad \frac{\partial L}{\partial v_2^2} = (\lambda + 2\mu)v_2^2 + (\lambda+\mu)v_1^1\,.
\end{aligned} \tag{7.40}$$

Let (X_1, X_2) be an integrable solution of (7.29) for this particular case. From (6.1) and (7.40), we have that, if

$$\begin{aligned}
\phi\colon \quad \mathbb{R}^2 &\to \mathbb{R}^2 \\
(x^1, x^2) &\mapsto (\phi^1(x), \phi^2(x))
\end{aligned}$$

is a solution of (X_1, X_2), then ϕ satisfies

$$\begin{aligned}
(\lambda + 2\mu)\partial_{11}\phi^1 + (\lambda+\mu)\partial_{12}\phi^2 + \mu\partial_{22}\phi^1 = 0\,, \\
\mu\partial_{11}\phi^2 + (\lambda+\mu)\partial_{12}\phi^1 + (\lambda+2\mu)\partial_{22}\phi^2 = 0\,,
\end{aligned}$$

which are *Navier's equations*, see [Olver (1986, 2007)]. These are the equations of motion for a viscous fluid in which one considers the effects of attraction and repulsion between neighboring molecules. Here λ and μ are coefficients of viscosity.

7.8 Equation of minimal surfaces

We consider again $Q = \mathbb{R}$ and (X_1, X_2) a solution of (7.29) where L is now the regular Lagrangian

$$\begin{aligned}
L\colon\ T\mathbb{R} \oplus T\mathbb{R} &\to \mathbb{R} \\
(q, v_1, v_2) &\mapsto \sqrt{1 + v_1^2 + v_2^2}\,.
\end{aligned}$$

Then one obtains,

$$\frac{\partial L}{\partial q} = 0\,, \quad \frac{\partial L}{\partial v_1} = \frac{v_1}{\sqrt{1 + (v_1)^2 + (v_2)^2}} \quad \frac{\partial L}{\partial v_2} = \frac{v_2}{\sqrt{1 + (v_1)^2 + (v_2)^2}}\,. \tag{7.41}$$

From (6.1) and (7.41), we deduce that if ϕ is solution of the 2-vector field (X_1, X_2), then ϕ satisfies

$$0 = (1 + (\partial_2\phi)^2)\partial_{11}\phi - 2\partial_1\phi\,\partial_2\phi\,\partial_{12}\phi + (1 + (\partial_1\phi)^2)\partial_{22}\phi\,,$$

which is the *equation of minimal surfaces,* (see for instance [Echeverría-Enríquez and Muñoz-Lecanda (1992); Olver (2007)]).

Remark 7.6. An alternative Lagrangian for the equation of minimal surfaces is given by $L(q, v_1, v_2) = 1 + v_1^2 + v_2^2$.

7.9 The massive scalar field

The equation of a scalar field ϕ (for instance the gravitational field) which acts on the 4-dimensional space-time is (see [Godstein, Poole Jr. and Safko (2001); Kijowski and Tulczyjew (1979)]):

$$(\Box + m^2)\phi = F'(\phi)\,, \tag{7.42}$$

where m is the mass of the particle over which the field acts, F is a scalar function such that $F(\phi) - \frac{1}{2}m^2\phi^2$ is the potential energy of the particle of mass m, and $\Box$ is the Laplace-Beltrami operator given by

$$\Box\phi\colon = div\, grad\phi = \frac{1}{\sqrt{-g}}\frac{\partial}{\partial x^\alpha}\left(\sqrt{-g}g^{\alpha\beta}\frac{\partial\phi}{\partial t^\beta}\right),$$

$(g_{\alpha\beta})$ being a pseudo-Riemannian metric tensor in the 4-dimensional space-time of signature $(- + ++)$, and $\sqrt{-g} = \sqrt{-\det g_{\alpha\beta}}$. In this case we suppose that the metric $(g_{\alpha\beta})$ is the Minkowski metric on $\mathbb{R}^4$, i.e.,

$$d(x^2)^2 + d(x^3)^2 + d(x^4)^2 - d(x^1)^2\,.$$

Consider the Hamiltonian function

$$\begin{aligned} H\colon \qquad & (T^1_4)^*\mathbb{R} \qquad \to \mathbb{R} \\ & (q, p^1, p^2, p^3, p^4) \mapsto \frac{1}{2}g_{\alpha\beta}p^\alpha p^\beta - \left(F(q) - \frac{1}{2}m^2q^2\right), \end{aligned}$$

where q denotes the scalar field ϕ and (q, p^1, p^2, p^3, p^4) the natural coordinates on $(T^1_4)^*\mathbb{R}$. Then

$$\frac{\partial H}{\partial q} = -\Big(F'(q) - m^2q\Big), \quad \frac{\partial H}{\partial p^\alpha} = g_{\alpha\beta}p^\beta\,. \tag{7.43}$$

Consider the 4-symplectic Hamiltonian equation

$$\iota_{X_1}\omega^1 + \iota_{X_2}\omega^2 + \iota_{X_3}\omega^3 + \iota_{X_4}\omega^4 = dH,$$

associated to the above Hamiltonian function. From (7.43) one obtains that, in natural coordinates, a 4-vector field solution of this equation has the following local expression

$$X_\alpha = g_{\alpha\beta}p^\beta\frac{\partial}{\partial q} + (X_\alpha)^\beta\frac{\partial}{\partial p^\beta}\,, \tag{7.44}$$

where the functions $(X_\alpha)^\beta \in \mathcal{C}^\infty((T^1_4)^*\mathbb{R})$ satisfies

$$F'(q) - m^2q = (X_1)^1 + (X_2)^2 + (X_3)^3 + (X_4)^4\,. \tag{7.45}$$

Let $\varphi\colon \mathbb{R}^4 \to (T^1_4)^*\mathbb{R}$, $\varphi(x) = (\psi(x), \psi^1(x), \psi^2(x), \psi^3(x), \psi^4(x))$ be an integral section of a 4-vector field solution of the 4-symplectic Hamiltonian equation. Then from (7.44) and (7.45) one obtains

$$\frac{\partial \psi}{\partial x^\alpha} = g_{\alpha\beta}\psi^\beta$$
$$F'(\psi(x)) - m^2\psi(x) = \frac{\partial \psi^1}{\partial x^1} + \frac{\partial \psi^2}{\partial x^2} + \frac{\partial \psi^3}{\partial x^3} + \frac{\partial \psi^4}{\partial x^4}\,.$$

Therefore, $\psi\colon \mathbb{R}^4 \to \mathbb{R}$ is a solution of the equation

$$F'(\psi) - m^2\psi = \frac{\partial}{\partial x^\alpha}\left(g^{\alpha\beta}\frac{\partial \psi}{\partial x^\beta}\right),$$

that is, ψ is a solution of the scalar field equation.

Remark 7.7. The scalar equation can be described using the Lagrangian approach with Lagrangian function

$$L(x^1, \dots, x^4, q, v_1, \dots, v_4) = \sqrt{-g}\Big(F(q) - \frac{1}{2}m^2q^2\Big) + \frac{1}{2}g^{\alpha\beta}v_\alpha v_\beta\,, \quad (7.46)$$

where q denotes the scalar field ϕ and v_α the partial derivative $\partial\phi/\partial x^\alpha$. Then equation (7.42) is the Euler-Lagrange equation associated to L. ◇

Remark 7.8. Some particular cases of the scalar field equation are the following:

(1) If $F = 0$ we obtain the linear scalar field equation.
(2) If $F(q) = m^2q^2$, we obtain the Klein-Gordon equation [José and Saletan (1998)]

$$(\Box + m^2)\psi = 0\,.$$

◇

PART 3

k-cosymplectic formulation of classical field theories

Part 2 of this book has been devoted to give a geometric description of certain kinds of classical field theories. The purpose of Part 3 is to extend the above study to classical field theories involving the independent parameters, i.e., the "space-time" coordinates $(x^1, \ldots, x^k)$ in an explicit way. In others words, in this part we shall give a geometrical description of classical field theories whose Lagrangian and Hamiltonian functions are of the form $L = L(x^\alpha, q^i, v^i_\alpha)$ and $H = H(x^\alpha, q^i, p^\alpha_i)$.

The model of the convenient geometrical structure for our approach is extracted of the so-called stable cotangent bundle of k^1-covelocities $\mathbb{R}^k \times (T^1_k)^*Q$. These structures are called k-cosymplectic manifolds and they were introduced by M. de León *et al.* [de León, Merino, Oubiña, Rodrigues and Salgado (1998); de León, Merino and Salgado (2001)].

In chapter 8 we shall recall the notion of k-cosymplectic manifold using as model $\mathbb{R}^k \times (T^1_k)^*Q$. Later, in chapter 9 we shall describe the k-cosymplectic formalism. This formulation can be applied to give a geometric version of the Hamilton-De Donder-Weyl and Euler-Lagrange equations for field theories. We also present several physical examples which can be described using this approach and the relationship between the Hamiltonian and Lagrange approaches.

Chapter 8

k-cosymplectic Geometry

The k-cosymplectic formulation is based on the so-called k-cosymplectic geometry. In this chapter we introduce these structures which are a generalization of the notion of cosymplectic forms.

Firstly, we describe the model of the k-cosymplectic manifolds, that is the stable cotangent bundle of k^1-covelocities $\mathbb{R}^k \times (T^1_k)^*Q$ and introduce the canonical structures living there. Using this model we introduce the notion of k-cosymplectic manifold. A complete description of these structures can be found in [de León, Merino, Oubiña, Rodrigues and Salgado (1998); de León, Merino and Salgado (2001)].

8.1 The stable cotangent bundle of k^1-covelocities $\mathbb{R}^k \times (T^1_k)^*Q$

Let $J^1(Q, \mathbb{R}^k)_0$ be the manifold of 1-jets of maps from Q to $\mathbb{R}^k$ with target at $0 \in \mathbb{R}^k$, which we described in Remark 2.1. Let us remember that this manifold is diffeomorphic to the cotangent bundle of k^1-covelocities $(T^1_k)^*Q$ via the diffeomorphism described in (2.5).

Indeed, for each $x \in \mathbb{R}^k$ we can consider the manifold $J^1(Q, \mathbb{R}^k)_x$ of 1-jets of maps from Q to $\mathbb{R}^k$ with target at $x \in \mathbb{R}^k$, i.e.,

$$J^1(Q,\mathbb{R}^k)_x = \bigcup_{q\in Q} J^1_{q,\,x}(Q,\mathbb{R}^k) = \bigcup_{q\in Q} \{j^1_{q,x}\sigma \,|\, \sigma : Q \to \mathbb{R}^k \text{ smooth}, \sigma(q) = x\}\,.$$

If we consider the collection of all these spaces, we obtain the set $J^1(Q, \mathbb{R}^k)$ of 1-jets of maps from Q to $\mathbb{R}^k$, i.e.,

$$J^1(Q,\mathbb{R}^k) = \bigcup_{x\in\mathbb{R}^k} J^1(Q,\mathbb{R}^k)_x\,.$$

The set $J^1(Q,\mathbb{R}^k)$ can be identified with $\mathbb{R}^k \times (T^1_k)^*Q$ via

$$\begin{array}{ccccc} J^1(Q,\mathbb{R}^k) & \to & \mathbb{R}^k \times J^1(Q,\mathbb{R}^k)_0 & \to & \mathbb{R}^k \times (T^1_k)^*Q \\ j^1_{q,x}\sigma & \to & (x, j^1_{q,0}\sigma_q) & \to & (x, d\sigma^1_q(q), \dots, d\sigma^k_q(q)) \end{array}, \tag{8.1}$$

where the last identification was described in (2.5), being $\sigma_q \colon Q \to \mathbb{R}^k$ the map defined by $\sigma_q(\tilde{q}) = \sigma(\tilde{q}) - \sigma(q)$ for any $\tilde{q} \in Q$.

Remark 8.1. We recall that the manifold of 1-jets of mappings from Q to $\mathbb{R}^k$, can be identified with the manifold $J^1\pi_Q$ of 1-jets of sections of the trivial bundle $\pi_Q \colon \mathbb{R}^k \times Q \to Q$, (a full description of the first-order jet bundle associated to an arbitrary bundle $E \to M$ can be found in [Saunders (1989)]). The diffeomorphism which establishes this relation is given by

$$\begin{array}{ccccc} J^1\pi_Q & \to & J^1(Q,\mathbb{R}^k) & \to & \mathbb{R}^k \times J^1(Q,\mathbb{R}^k)_0 \\ j^1_q\phi & \to & j^1_{q,\sigma(q)}\sigma & \to & (\sigma(q), j^1_{q,0}\sigma_q) \end{array}$$

where $\phi \colon Q \to \mathbb{R}^k \times Q$ is a section of π_Q, $\sigma \colon Q \to \mathbb{R}^k$ is given by $\sigma = \pi_{\mathbb{R}^k} \circ \phi$ being $\pi_{\mathbb{R}^k} \colon \mathbb{R}^k \times Q \to \mathbb{R}^k$ the canonical projection and $\sigma_q \colon Q \to \mathbb{R}^k$ is defined by $\sigma_q(\tilde{q}) = \sigma(\tilde{q}) - \sigma(q)$ for any $\tilde{q} \in Q$. $\diamond$

From the above comments we know that an element of $J^1(Q,\mathbb{R}^k) \equiv \mathbb{R}^k \times (T^1_k)^*Q$ is a $(q+1)$-tuple $(x, \nu_{1_q}, \dots, \nu_{k_q})$ where $x \in \mathbb{R}^k$ and $(\nu_{1_q}, \dots, \nu_{k_q}) \in (T^1_k)^*Q$. Thus we can consider the following canonical projections:

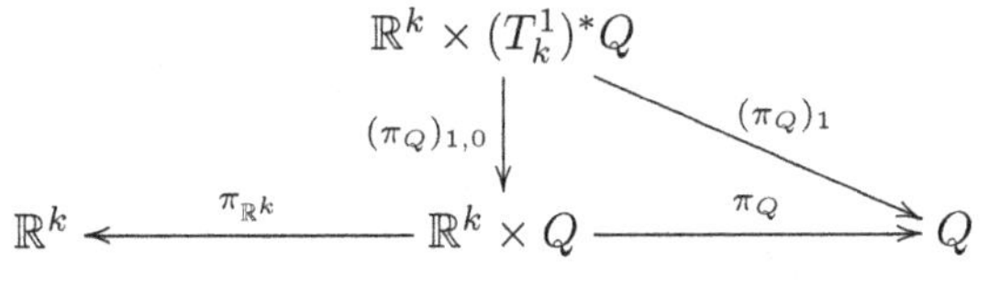

$$\begin{array}{ccccc} & & \mathbb{R}^k \times (T^1_k)^*Q & & \\ & & \downarrow{\scriptstyle (\pi_Q)_{1,0}} & \searrow^{(\pi_Q)_1} & \\ \mathbb{R}^k & \xleftarrow{\pi_{\mathbb{R}^k}} & \mathbb{R}^k \times Q & \xrightarrow{\pi_Q} & Q \end{array}$$

defined by

$$\begin{array}{ll} (\pi_Q)_{1,0}(x, \nu_{1_q}, \dots, \nu_{k_q}) = (x,q), & \pi_{\mathbb{R}^k}(x,q) = x, \\ (\pi_Q)_1(x, \nu_{1_q}, \dots, \nu_{k_q}) = q, & \pi_Q(x,q) = q, \end{array} \tag{8.2}$$

with $x \in \mathbb{R}^k$, $q \in Q$ and $(\nu_{1_q}, \dots, \nu_{k_q}) \in (T^1_k)^*Q$.

In the following diagram we collect the notation used for the projections in this part of the book:

If (q^i) with $1 \leq i \leq n$, is a local coordinate system defined on an open set $U \subset Q$, the induced ***local coordinates*** $(x^\alpha, q^i, p^\alpha_i)$, $1 \leq i \leq n$, $1 \leq \alpha \leq k$

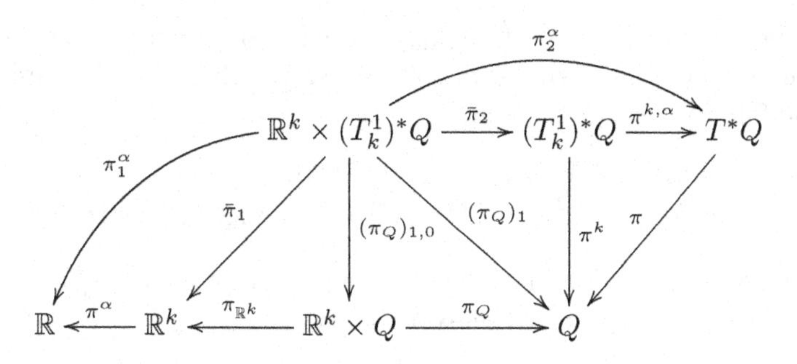

Fig. 8.1 Canonical projections associated to $\mathbb{R}^k \times (T_k^1)^*Q$

on $\mathbb{R}^k \times (T_k^1)^*U = \Big((\pi_Q)_1\Big)^{-1}(U)$ are given by

$$\begin{aligned} x^\alpha(x, \nu_{1q}, \ldots, \nu_{kq}) &= x^\alpha(x) = x^\alpha\,, \\ q^i(x, \nu_{1q}, \ldots, \nu_{kq}) &= q^i(q)\,, \\ p_i^\alpha(x, \nu_{1q}, \ldots, \nu_{kq}) &= \nu_{\alpha q}\left(\frac{\partial}{\partial q^i}\Big|_q\right). \end{aligned} \tag{8.3}$$

Thus, $\mathbb{R}^k \times (T_k^1)^*Q$ is endowed with a structure of differentiable manifold of dimension $k + n(k+1)$, and the manifold $\mathbb{R}^k \times (T_k^1)^*Q$ with the projection $(\pi_Q)_1$ has the structure of a vector bundle over Q.

If we consider the identification (8.1), the above coordinates can be defined in terms of 1-jets of map from Q to $\mathbb{R}^k$ in the following way

$$x^\alpha(j^1_{q,x}\sigma) = x^\alpha(x) = x^\alpha\,, \quad q^i(j^1_{q,x}\sigma) = q^i(q)\,, \quad p^i_\alpha(j^1_{q,x}\sigma) = \frac{\partial \sigma^\alpha}{\partial q^i}\Big|_q\,.$$

On $\mathbb{R}^k \times (T_k^1)^*Q$ we can define a family of canonical forms as follows

$$\eta^\alpha = (\pi_1^\alpha)^* dx, \quad \Theta^\alpha = (\pi_2^\alpha)^*\theta \quad \text{and} \quad \Omega^\alpha = (\pi_2^\alpha)^*\omega\,, \tag{8.4}$$

with $1 \leq \alpha \leq k$, being $\pi_1^\alpha : \mathbb{R}^k \times (T_k^1)^*Q \to \mathbb{R}$ and $\pi_2^\alpha : \mathbb{R}^k \times (T_k^1)^*Q \to T^*Q$ the projections defined by

$$\pi_1^\alpha(x, \nu_{1q}, \ldots, \nu_{kq}) = x^\alpha\,, \quad \pi_2^\alpha(x, \nu_{1q}, \ldots, \nu_{kq}) = \nu_{\alpha_q}$$

and θ and ω the canonical Liouville and symplectic forms on T^*Q, respectively. Let us observe that, since $\omega = -d\theta$, then $\Omega^\alpha = -d\Theta^\alpha$.

If we consider a local coordinate system $(x^\alpha, q^i, p_i^\alpha)$ on $\mathbb{R}^k \times (T_k^1)^*Q$ (see (8.3)), the ***canonical forms*** η^α, Θ^α and Ω^α have the following local expressions:

$$\eta^\alpha = dx^\alpha, \quad \Theta^\alpha = p_i^\alpha dq^i\,, \quad \Omega^\alpha = dq^i \wedge dp_i^\alpha\,. \tag{8.5}$$

Moreover, let $\mathcal{V}^* = \ker\left((\pi_Q)_{1,0}\right)_*$; then a simple inspection in local coordinates shows that the forms η^α and Ω^α, with $1 \leq \alpha \leq k$ are closed and the following relations hold:

(1) $dx^1 \wedge \cdots \wedge dx^k \neq 0, \quad dx^\alpha|_{\mathcal{V}^*} = 0, \quad \Omega^\alpha|_{\mathcal{V}^* \times \mathcal{V}^*} = 0,$
(2) $(\cap_{\alpha=1}^k \ker dx^\alpha) \cap (\cap_{\alpha=1}^k \ker \Omega^\alpha) = \{0\}, \quad \dim(\cap_{\alpha=1}^k \ker \Omega^\alpha) = k.$

Remark 8.2. Let us observe that the canonical forms on $(T_k^1)^*Q$ and on $\mathbb{R}^k \times (T_k^1)^*Q$ are related by the expressions

$$\theta^\alpha = (\bar{\pi}_2)^*\theta^\alpha \quad \text{and} \quad \Omega^\alpha = (\bar{\pi}_2)^*\omega^\alpha, \tag{8.6}$$

with $1 \leq \alpha \leq k$. ◇

8.2 k-cosymplectic geometry

From the above model, that is, the stable cotangent bundle of k^1-covelocities with the canonical forms (8.4), M. de León and collaborators have introduced the notion of k-cosymplectic structures in [de León, Merino, Oubiña, Rodrigues and Salgado (1998); de León, Merino and Salgado (2001)].

Let us recall that the k-cosymplectic manifolds provide a natural arena to develop classical field theories as an alternative to other geometrical settings as the polysymplectic geometry [Giachetta, Mangiarotti and Sardanashvily (1997, 1999); Sardanashvily (1993, 1996)] or multisymplectic geometry.

Before introducing the formal definition of k-cosymplectic manifold we consider the linear case.

8.2.1 *k-cosymplectic vector spaces*

Inspired by the above geometrical model one can define k-cosymplectic structures on a vector space in the following way (see [Merino (1997)]).

Definition 8.1. Let E be a $k + n(k+1)$-dimensional vector space. A family $(\eta^\alpha, \Omega^\alpha, V; 1 \leq \alpha \leq k)$ be where $\eta^1, \dots, \eta^k$ are 1-forms, $\Omega^1, \dots, \Omega^k$ are 2-forms and V is a vector subspace of E of dimension nk, defines a ***k-cosymplectic structure*** on the vector space E if the following conditions hold:

(1) $\eta^1 \wedge \ldots \wedge \eta^k \neq 0,$

(2) $\dim(\ker\Omega^1 \cap \ldots \cap \ker\Omega^k) = k$,
(3) $(\cap_{\alpha=1}^k \ker\eta^\alpha) \cap (\cap_{\alpha=1}^k \ker\Omega^\alpha) = \{0\}$,
(4) $\eta^\alpha|_V = 0$, $\Omega^\alpha|_{V\times V} = 0$, $1 \leq \alpha \leq k$.

$(E, \eta^\alpha, \Omega^\alpha, V)$ is called ***k-cosymplectic vector space***.

Remark 8.3. If $k = 1$, then E is a vector space of dimension $2n + 1$ and we have a family (η, Ω, V) given by a 1-form η, a 2-form Ω and a subspace $V \subset E$ of dimension n.

From conditions (2) and (3) of the above definition one deduces that $\eta \wedge \Omega^n \neq 0$ since $\dim\ker\omega = 1$, and then $\operatorname{rank}\Omega = 2n$, moreover $\ker\eta \cap \ker\Omega = 0$.

The pair (η, Ω) define a cosymplectic structure on E. Moreover, from condition (4) one deduce that (η, Ω, V) define a stable almost cotangent structure on E. ◇

Given a k-cosymplectic structure on a vector space one can prove the following results (see [Merino (1997)]):

Theorem 8.1 (Darboux coordinates). *If $(\eta^\alpha, \Omega^\alpha, V; 1 \leq \alpha \leq k)$ is a k-cosymplectic structure on E then there exists a basis $(\eta^1, \ldots, \eta^k, \gamma^i, \gamma_i^\alpha; 1 \leq i \leq n, 1 \leq \alpha \leq k)$ of E^* such that*

$$\Omega^\alpha = \gamma^i \wedge \gamma_i^\alpha .$$

For every k-cosymplectic structure $(\eta^\alpha, \Omega^\alpha, V; 1 \leq \alpha \leq k)$ on a vector space E, there exists a family of k vectors $R_1, \ldots, R_k$, which are called ***Reeb vectors***, characterized by the conditions

$$\iota_{R_\alpha}\eta^\beta = \delta_\alpha^\beta, \; \iota_{R_\alpha}\omega^\beta = 0 .$$

8.2.2 *k-cosymplectic manifolds*

We turn now to the globalization of the ideas of the previous section to manifolds. The following definition was introduced in [de León, Merino, Oubiña, Rodrigues and Salgado (1998)]:

Definition 8.2. Let M be a differentiable manifold of dimension $k(n+1)+n$. A ***k-cosymplectic structure*** on M is a family $(\eta^\alpha, \Omega^\alpha, V; 1 \leq \alpha \leq k)$, where each η^α is a closed 1-form, each Ω^α is a closed 2-form and V is an integrable nk-dimensional distribution on M satisfying

(1) $\eta^1 \wedge \cdots \wedge \eta^k \neq 0$, $\quad \eta^\alpha|_V = 0$, $\quad \Omega^\alpha|_{V\times V} = 0$,

(2) $(\cap_{\alpha=1}^{k} \ker \eta^\alpha) \cap (\cap_{\alpha=1}^{k} \ker \Omega^\alpha) = \{0\}, \quad \dim (\cap_{\alpha=1}^{k} \ker \Omega^\alpha) = k.$

M is said to be a ***k*-cosymplectic manifold.**

In particular, if $k = 1$, then $dim\, M = 2n+1$ and (η, Ω) is a cosymplectic structure on M.

For every k-cosymplectic structure $(\eta^\alpha, \Omega^\alpha, \mathcal{V})$ on M, there exists a family of k vector fields $\{R_\alpha\}$, which are called ***Reeb vector fields***, characterized by the following conditions

$$\iota_{R_\alpha} \eta^\beta = \delta^\beta_\alpha \quad , \quad \iota_{R_\alpha} \Omega^\beta = 0 \; .$$

In the canonical model $R_\alpha = \dfrac{\partial}{\partial x^\alpha}$.

The following theorem has been proved in [de León, Merino, Oubiña, Rodrigues and Salgado (1998)]:

Theorem 8.2 (Darboux Theorem). *If M is a k-cosymplectic manifold, then around each point of M there exist local coordinates $(x^\alpha, q^i, p^\alpha_i; 1 \leq A \leq k, 1 \leq i < n)$ such that*

$$\eta^\alpha = dx^\alpha, \quad \Omega^\alpha = dq^i \wedge dp^\alpha_i, \quad V = \left\langle \frac{\partial}{\partial p^1_i}, \dots, \frac{\partial}{\partial p^k_i} \right\rangle_{i=1,\dots,n} .$$

The canonical model for these geometrical structures is

$$(\mathbb{R}^k \times (T^1_k)^* Q, \eta^\alpha, \Omega^\alpha, V^*).$$

Example 8.1. Let (N, ω^α, V) be an arbitrary k-symplectic manifold. Then, denoting by

$$\pi_{\mathbb{R}^k} : \mathbb{R}^k \times N \to \mathbb{R}^k, \qquad \pi_N : \mathbb{R}^k \times N \to N$$

the canonical projections, we consider the differential forms

$$\eta^\alpha = \pi^*_{\mathbb{R}^k}(dx^\alpha), \quad \Omega^\alpha = \pi^*_N \omega^\alpha \, ,$$

and the distribution V in N defines a distribution $\mathcal{V}$ in $M = \mathbb{R}^k \times N$ in a natural way. All conditions given in Definition 8.2 are verified, and hence $\mathbb{R}^k \times N$ is a k-cosymplectic manifold.

Chapter 9

k-cosymplectic Formalism

In this chapter we describe the k-cosymplectic formalism. As we shall see in the following chapters, using this formalism we can study classical field theories that explicitly involve the space-time coordinates on the Hamiltonian and Lagrangian. This is the principal difference with the k-symplectic approach. As in previous case, in this formalism the notion of k-vector field is fundamental; let us recall that this notion was introduced in section 3.1 for an arbitrary manifold M.

9.1 k-cosymplectic Hamiltonian equations

Let $(M, \eta^\alpha, \Omega^\alpha, V)$ be a k-cosymplectic manifold and H a Hamiltonian on M, that is, a function $H\colon M \to \mathbb{R}$ defined on M.

Definition 9.1. The family $(M, \eta^\alpha, \Omega^\alpha, H)$ is called ***k-cosymplectic Hamiltonian system.***

Let $(M, \eta^\alpha, \Omega^\alpha, H)$ be a k-cosymplectic Hamiltonian system and $\mathbf{X} = (X_1, \dots, X_k)$ a k-vector field on M solution of the system of equations

$$\begin{aligned} \eta^\alpha(X_\beta) &= \delta^\alpha_\beta, \quad 1 \leq \alpha, \beta \leq k \\ \sum_{\alpha=1}^{k} \iota_{X_\alpha}\Omega^\alpha &= dH - \sum_{\alpha=1}^{k} R_\alpha(H)\eta^\alpha , \end{aligned} \tag{9.1}$$

where $R_1, \dots, R_k$ are the Reeb vector fields associated with the k-cosymplectic structure on M.

Given a local coordinate system $(x^\alpha, q^i, p^\alpha_i)$, each X_α, $1 \leq \alpha \leq k$ is locally given by

$$X_\alpha = (X_\alpha)_\beta \frac{\partial}{\partial x^\alpha} + (X_\alpha)^i \frac{\partial}{\partial q^i} + (X_\alpha)^\beta_i \frac{\partial}{\partial v^i_\beta} \,.$$

Now, since

$$dH = \frac{\partial H}{\partial x^\alpha} dx^\alpha + \frac{\partial H}{\partial q^i} dq^i + \frac{\partial H}{\partial p_i^\alpha} dp_i^\alpha$$

and

$$\eta^\alpha = dx^\alpha, \quad \omega^\alpha = dq^i \wedge dq^i, \quad R_\alpha = \partial/\partial x^\alpha$$

we deduce that equation (9.1) is locally equivalent to the following conditions

$$(X_\alpha)_\beta = \delta_\alpha^\beta, \quad \frac{\partial H}{\partial q^i} = -\sum_{\beta=1}^k (X_\beta)_i^\beta \quad \frac{\partial H}{\partial p_i^\alpha} = (X_\alpha)^i \tag{9.2}$$

with $1 \leq i \leq n$ and $1 \leq \alpha \leq k$.

Let us suppose that $\mathbf{X} = (X_1, \ldots, X_k)$ is integrable, and

$$\varphi : : \mathbb{R}^k \to M$$

$$x \quad \to \varphi(x) = (\psi_\alpha(x), \psi^i(x), \psi_i^\alpha(x))$$

is an integral section of $\mathbf{X}$, then

$$\varphi_*(x)\Big(\frac{\partial}{\partial x^\alpha}\Big|_x\Big) = \frac{\partial \psi_\beta}{\partial x^\alpha}\Big|_x \frac{\partial}{\partial x^\beta}\Big|_{\varphi(x)} + \frac{\partial \psi^i}{\partial x^\alpha}\Big|_x \frac{\partial}{\partial q^i}\Big|_{\varphi(x)} + \frac{\partial \psi_i^\beta}{\partial x^\alpha}\Big|_x \frac{\partial}{\partial p_i^\beta}\Big|_{\varphi(x)}. \tag{9.3}$$

From (9.3) we obtain that φ is given by $\varphi(x) = (x, \psi^i(x), \psi_i^\alpha(x))$ and the following equations

$$\frac{\partial \psi_\beta}{\partial x^\alpha}\Big|_x = \delta_\beta^\alpha, \quad \frac{\partial \psi^i}{\partial x^\alpha}\Big|_x = (X_\alpha)^i(\varphi(x)), \quad \frac{\partial \psi_i^\beta}{\partial x^\alpha}\Big|_x = (X_\alpha)_i^\beta(\varphi(x)), \tag{9.4}$$

hold.

This theory can be summarized in the following

Theorem 9.1. *Let $(M, \eta^\alpha, \Omega^\alpha, H)$ a k-cosymplectic Hamiltonian system and $\mathbf{X} = (X_1, \ldots, X_k)$ a k-vector field on M solution of the system of equations*

$$\eta^\alpha(X_\beta) = \delta_\beta^\alpha, \quad 1 \leq \alpha, \beta \leq k$$

$$\sum_{\alpha=1}^k \iota_{X_\alpha} \Omega^\alpha = dH - \sum_{\alpha=1}^k R_\alpha(H)\eta^\alpha\,,$$

where $R_1, \ldots, R_k$ are the Reeb vector fields associated with the k-cosymplectic structure on M.

If $\mathbf{X}$ is integrable and $\varphi : \mathbb{R}^k \to M$, $\varphi(x) = (x^\alpha, \psi^i(x), \psi_i^\alpha(x))$ is an integral section of the k-vector field $\mathbf{X}$, then φ is a solution of the following system of partial differential equations

$$\frac{\partial H}{\partial q^i}\Big|_{\varphi(x)} = -\sum_{\alpha=1}^k \frac{\partial \psi_i^\alpha}{\partial x^\alpha}\Big|_t, \quad \frac{\partial H}{\partial p_i^\alpha}\Big|_{\varphi(x)} = \frac{\partial \psi^i}{\partial x^\alpha}\Big|_x.$$

□

From now on, we shall call these equations (9.1) as ***k-cosymplectic Hamiltonian equations.***

Definition 9.2. A k-vector field $\mathbf{X} = (X_1, \dots, X_k) \in \mathfrak{X}^k(M)$ is called a ***k-cosymplectic Hamiltonian k-vector field*** for a k-cosymplectic Hamiltonian system $(M, \eta^\alpha, \Omega^\alpha, H)$ if $\mathbf{X}$ is a solution of (9.1). We denote by $\mathfrak{X}^k_H(M)$ the set of all k-cosymplectic Hamiltonian k-vector fields.

It should be noticed that equations (9.1) always have a solution but this is not unique. In fact, if $(M, \eta^\alpha, \Omega^\alpha, V)$ is a k-cosymplectic manifold we can define two vector bundle morphism $\Omega^\flat \colon TM \to (T^1_k)^*M$ and $\Omega^\sharp \colon T^1_kM \to T^*M$ as follows:

$$\Omega^\flat(X) = (\iota_X\Omega^1 + \eta^1(X)\eta^1, \dots, \iota_X\Omega^k + \eta^k(X)\eta^k)$$

and $\Omega^\sharp(X_1, \dots, X_k)$ such that

$$\Omega^\sharp(X_1, \dots, X_k)(Y) = \mathrm{trace}((\Omega^\flat(X_\beta))_\alpha(Y)) = \sum_{\alpha=1}^k ((\Omega^\flat(X_\alpha))_\alpha(Y))$$

for all $Y \in TM$, i.e.,

$$\Omega^\sharp(X_1, \dots, X_k) = \sum_{\alpha=1}^k (\iota_{X_\alpha}\Omega^\alpha + \eta^\alpha(X_\alpha)\eta^\alpha)\,.$$

The above morphisms induce two morphisms of $\mathcal{C}^\infty(M)$-module between the corresponding spaces of sections. Let us observe that the equations (9.1) are equivalent to

$$\eta^\alpha(X_\beta) = \delta^\alpha_\beta, \quad \forall \alpha, \beta,$$
$$\Omega^\sharp(X_1, \dots, X_k) = dH + \sum_{\alpha=1}^k (1 - R_\alpha(H))\eta^\alpha\,,$$

where $R_1, \dots, R_k$ are the Reeb vector fields of the k-cosymplectic structure $(\eta^\alpha, \Omega^\alpha, V)$.

Remark 9.1. If $k = 1$ then $\Omega^\flat = \Omega^\sharp$ is defined from TM to T^*M, and it is the morphism $\chi_{\eta,\Omega}$ associated to the cosymplectic manifold (M, η, Ω) and defined by

$$\chi_{\eta,\Omega}(X) = \iota_X\Omega + \eta(X)\eta\,,$$

(for more details see [Albert (1989); Cantrijn, de León and Lacomba (1992)] and Appendix B). ◇

Next we shall discuss the existence of solution of the above equations. From the local conditions (9.2) we can define a k-vector field that satisfies (9.2), on a neighborhood of each point $x \in M$. For example, we can put

$$(X_\alpha)^\beta = \delta^\beta_\alpha\,, \quad (X_1)^1_i = \frac{\partial H}{\partial q^i}\,, \quad (X_\alpha)^\beta_i = 0 \text{ for } \alpha \neq 1 \neq \beta\,, \quad (X_\alpha)^i = \frac{\partial H}{\partial p^A_i}\,.$$

Now by using a partition of unity in the manifold M, one can construct a global k-vector field which is a solution of (9.1), (see [de León, Merino, Oubiña, Rodrigues and Salgado (1998)]).

Equations (9.1) have no, in general, unique solution. In fact, denoting by $\mathcal{M}_k(C^\infty(M))$ the space of matrices of order k whose entries are functions on M, we define the vector bundle morphism

$$\begin{aligned} \eta^\sharp : \quad & T^1_k M \longrightarrow \mathcal{M}_k(C^\infty(M)) \\ & (X_1, \dots, X_k) \mapsto \eta^\sharp(X_1, \dots, X_k) = (\eta_\alpha(X_\beta))\,. \end{aligned} \tag{9.5}$$

Then the solutions of (9.1) are given by $(X_1, \dots, X_k) + (\ker \Omega^\sharp \cap \ker \eta^\sharp)$, where $(X_1, \dots, X_k)$ is a particular solution.

Let us observe that given a k-vector field $Y = (Y_1, \dots, Y_k)$ the condition $Y \in \ker \Omega^\sharp \cap \ker \eta^\sharp$ is locally equivalent to the conditions

$$(Y_\beta)_\alpha = 0, \quad Y^i_\beta = 0, \quad \sum_{\alpha=1}^{k} (Y_\alpha)^\alpha_i = 0\,. \tag{9.6}$$

Finally, in the proof of Theorem 9.1 it is necessary to assume the integrability of the k-vector field $(X_1, \dots, X_k)$, and since the k vector fields $X_1, \dots, X_k$ on M are linearly independent, the integrability condition is equivalent to require that $[X_\alpha, X_\beta] = 0$, for all $1 \leq \alpha, \beta \leq k$.

Remark 9.2. Sometimes the Hamiltonian (or Lagrangian) functions are not defined on a k-cosymplectic manifold, for instance, in the reduction theory where the reduced "phase spaces" are not, in general, k-cosymplectic manifolds, even when the original phase space is a k-cosymplectic manifold. In mechanics this problem is solved using Lie algebroids (see [de León, Marrero and Martínez (2005); Martínez (2001,b); Weinstein (1996)]). In [Martín de Diego and Vilariño (2010)] we introduce a geometric description of classical field theories on Lie algebroids in the frameworks of k-cosymplectic geometry. classical field theories on Lie algebroids have already been studied in the literature. For instance, the multisymplectic formalism on Lie algebroids was presented in [Martínez (2004, 2005)], the k-symplectic formalism on Lie algebroids in [de León, Martín de Diego, Salgado and Vilariño (2009)], in [Vankerschaver and Cantrijn (2007)] a geometric framework for discrete field theories on Lie groupoids has been discussed. ◇

9.2 Example: massive scalar field

Consider the 4-cosymplectic Hamiltonian equation

$$
\begin{aligned}
& dx^\alpha(X_\beta) = \delta_{\alpha\beta}, \quad 1 \leq \alpha, \beta \leq 4 \\
& \sum_{\alpha=1}^{4} \iota_{X_\alpha}\Omega^\alpha = dH - \sum_{\alpha=1}^{4} R_\alpha(H)dx^\alpha
\end{aligned}
\tag{9.7}
$$

associated to the Hamiltonian function $H \in \mathcal{C}^\infty(\mathbb{R}^4 \times (T^1_4)^*\mathbb{R})$,

$$
H(x^1,x^2,x^3,x^4,q,p^1,p^2,p^3,p^4) = \frac{1}{2\sqrt{-g}} g_{\alpha\beta}p^\alpha p^\beta - \sqrt{-g}\left(F(q) - \frac{1}{2}m^2q^2\right).
$$

If (X_1, X_2, X_3, X_4) is a solution of (9.7), then from the following identities

$$
\frac{\partial H}{\partial q} = -\sqrt{-g}\Big(F'(q) - m^2 q\Big), \quad \frac{\partial H}{\partial p^\alpha} = \frac{1}{\sqrt{-g}} g_{\alpha\beta}p^\beta \tag{9.8}
$$

and from (9.2) we obtain, in natural coordinates, the local expression of each X_α:

$$
X_\alpha = \frac{\partial}{\partial x^\alpha} + \frac{1}{\sqrt{-g}} g_{\alpha\beta}p^\beta \frac{\partial}{\partial q} + (X_\alpha)^\beta \frac{\partial}{\partial p^\beta}, \tag{9.9}
$$

where the functions $(X_\alpha)^\beta \in \mathcal{C}^\infty(\mathbb{R}^4 \times (T^1_4)^*\mathbb{R})$ satisfies

$$
\sqrt{-g}\Big(F'(q) - m^2 q\Big) = (X_1)^1 + (X_2)^2 + (X_3)^3 + (X_4)^4. \tag{9.10}
$$

Let us suppose that $\mathbf{X} = (X_1, X_2, X_3, X_4)$ is integrable and $\varphi\colon \mathbb{R}^4 \to \mathbb{R}^4 \times (T^1_4)^*\mathbb{R}$, with

$$
\varphi(x) = (x, \psi(x), \psi^1(x), \psi^2(x), \psi^3(x), \psi^4(x))
$$

is an integral section of $\mathbf{X}$, then one obtains that $(\psi(x), \psi^1(x), \psi^2(x), \psi^3(x), \psi^4(x))$ are solution of the following equations

$$
\begin{aligned}
& \frac{\partial \psi}{\partial x^\alpha} = \frac{1}{\sqrt{-g}} g_{\alpha\beta}\psi^\beta \\
& \sqrt{-g}\Big(F'(\psi) - m^2\psi\Big) = \frac{\partial \psi^1}{\partial x^1} + \frac{\partial \psi^2}{\partial x^2} + \frac{\partial \psi^3}{\partial x^3} + \frac{\partial \psi^4}{\partial x^4}.
\end{aligned}
$$

Thus $\psi\colon \mathbb{R}^4 \to \mathbb{R}$ is a solution of the equation

$$
\sqrt{-g}\Big(F'(\psi) - m^2\psi\Big) = \sqrt{-g}\frac{\partial}{\partial x^\alpha}\left(g^{\alpha\beta}\frac{\partial \psi}{\partial t^\beta}\right),
$$

that is, ψ is a solution of the scalar field equation (for more details about this equation see sections 7.9 and 13.2).

Chapter 10

Hamiltonian Classical Field Theories

In this chapter we shall study Hamiltonian classical field theories when the Hamiltonian function involves the space-time coordinates, that is, H is a function defined on $\mathbb{R}^k \times (T^1_k)^*Q$. Therefore, we shall discuss the Hamilton-De Donder-Weyl equations (HDW equations for short) which have the following local expression

$$\left.\frac{\partial H}{\partial q^i}\right|_{\varphi(x)} = -\sum_{\alpha=1}^{k} \left.\frac{\partial \psi^\alpha_i}{\partial x^\alpha}\right|_t, \quad \left.\frac{\partial H}{\partial p^\alpha_i}\right|_{\varphi(x)} = \left.\frac{\partial \psi^i}{\partial x^\alpha}\right|_x. \tag{10.1}$$

A solution of these equations is a map

$$\begin{aligned} \varphi : \mathbb{R}^k &\longrightarrow \mathbb{R}^k \times (T^1_k)^*Q \\ x &\to \varphi(x) = (x^\alpha, \psi^i(x), \psi^\alpha_i(x)) \end{aligned}$$

where $1 \leq i \leq n$, $1 \leq \alpha \leq k$.

In the classical approach these equations can be obtained from a multiple integral variational problem. In this chapter we shall describe this variational approach and, then, we shall give a new geometric way of obtaining the HDW equations using the k-cosymplectic formalism described in chapter 9 when the k-cosymplectic manifolds is just the geometrical model, i.e., $(M = \mathbb{R}^k \times (T^1_k)^*Q, \eta^1, \dots, \eta^k, \Omega^1, \dots, \Omega^k, V)$ as it has been described in section 8.1.

10.1 Variational approach

In this section we shall see that the HDW field equations (10.1) are obtained from a variational principle on the space of smooth maps with compact support $\mathcal{C}^\infty_C(\mathbb{R}^k, \mathbb{R}^k \times (T^1_k)^*Q)$.

To describe this variational principle we need the notion of prolongation of diffeomorphism and vector fields from Q to $\mathbb{R}^k \times (T^1_k)^*Q$, which we shall introduce now. First, we shall describe some properties of the π_Q-projectable vector fields.

10.1.1 *Prolongation of vector fields*

On the manifold $\mathbb{R}^k \times (T^1_k)^*Q$ there exist two families or groups of vector fields that are relevant for our purposes. The first of these families is the set of vector fields which are obtained as canonical prolongations of vectors field on $\mathbb{R}^k \times Q$ to $\mathbb{R}^k \times (T^1_k)^*Q$. The second interesting family is the set of π_Q-projectable vector fields defined on $\mathbb{R}^k \times Q$. In this paragraph we briefly describe these two sets of vector fields.

Definition 10.1. Let Z be a vector field on $\mathbb{R}^k \times Q$. Z is say to be π_Q-***projectable*** if there exists a vector field $\bar{Z}$ on Q, such that

$$(\pi_Q)_* \circ Z = \bar{Z} \circ \pi_Q \,.$$

To find the coordinate representation of the vector field Z we use coordinates $(x^\alpha, q^i, \dot{x}^\alpha, \dot{q}^i)$ on $T(\mathbb{R}^k \times Q)$ and (x^α, q^i) on $\mathbb{R}^k \times Q$. Since Z is a section of $T(\mathbb{R}^k \times Q) \to \mathbb{R}^k \times Q$, the x^α and q^i components of the coordinate representation are fixed, so that Z is determined by the functions $Z_\alpha = \dot{x}^\alpha \circ Z$ and $Z^i = \dot{q}^i \circ Z$, i.e.,

$$Z(x,q) = Z_\alpha(x,q)\frac{\partial}{\partial x^\alpha}\Big|_{(x,q)} + Z^i(x,q)\frac{\partial}{\partial q^i}\Big|_{(x,q)} .$$

On the other hand, $\bar{Z} \in \mathfrak{X}(Q)$, is locally expressed by

$$\bar{Z}(q) = \bar{Z}^i(q)\frac{\partial}{\partial q^i}\Big|_q ,$$

where $\bar{Z}^i \in C^\infty(Q)$.

Now the condition of Definition 10.1 implies that

$$Z^i(x,q) = (\bar{Z}^i \circ \pi_Q)(x,q) = \bar{Z}^i(q) \,.$$

We usually write Z^i instead of $\bar{Z}^i$. With this notation we have

$$Z(x,q) = Z_\alpha(x,q)\frac{\partial}{\partial x^\alpha}\Big|_{(x,q)} + Z^i(q)\frac{\partial}{\partial q^i}\Big|_{(x,q)} ,$$

$$\bar{Z}(q) = Z^i(q)\frac{\partial}{\partial q^i}\Big|_q .$$

As a consequence, we deduce that if $\{\sigma_s\}$ is the one-parameter group of diffeomorphism associated to a π_Q-projectable vector field $Z \in \mathfrak{X}(\mathbb{R}^k \times Q)$ and $\{\bar{\sigma}_s\}$ is the one-parameter group of diffeomorphism associate to $\bar{Z}$, then

$$\bar{\sigma}_s \circ \pi_Q = \pi_Q \circ \sigma_s\,.$$

Given a π_Q-projectable vector field $Z \in \mathfrak{X}(\mathbb{R}^k \times Q)$, we can define a vector field Z^{1*} on $\mathbb{R}^k \times (T^1_k)^*Q$ such that it is $(\pi_Q)_{1,0}$-projectable and its projection is Z. Here we give the idea of the definition. A complete description of this notion can be found in [Saunders (1989)] where the author define the prolongation of vector fields to the first-order jet bundle of an arbitrary bundle $E \to M$.

Before to construct the prolongation of a vector field it is necessary the following definition:

Definition 10.2. Let $f\colon \mathbb{R}^k \times Q \to \mathbb{R}^k \times Q$ be a map and $\bar{f}\colon Q \to Q$ be a diffeomorphism, such that $\pi_Q \circ f = \bar{f} \circ \pi_Q$. The ***first prolongation*** of f is a map

$$j^{1\,*}f\colon J^1(Q,\mathbb{R}^k) \equiv \mathbb{R}^k \times (T^1_k)^*Q \to J^1(Q,\mathbb{R}^k) \equiv \mathbb{R}^k \times (T^1_k)^*Q$$

defined by

$$(j^{1\,*}f)(j_{q,\sigma(q)}\sigma) = j^1_{(\bar{f}(q),\tilde{\sigma}(\bar{f}(q)))}\tilde{\sigma} \tag{10.2}$$

where $\sigma\colon Q \to \mathbb{R}^k$, $j_{q,\sigma(q)}\sigma \in J^1(Q,\mathbb{R}^k)$ and $\tilde{\sigma}\colon Q \to \mathbb{R}^k$ is the map given by the composition,

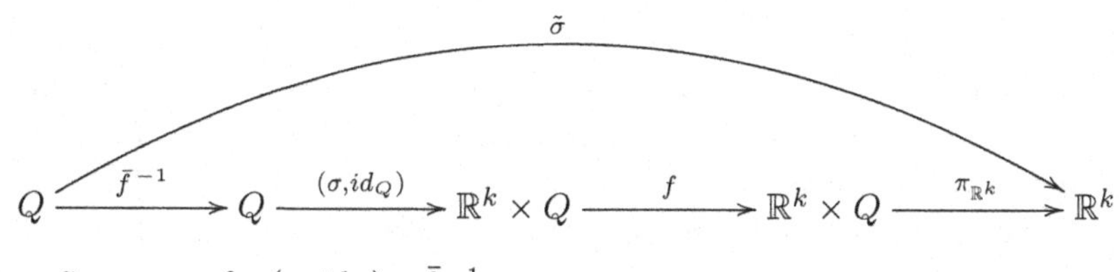

i.e., $\tilde{\sigma} = \pi_{\mathbb{R}^k} \circ f \circ (\sigma, id_Q) \circ \bar{f}^{-1}$.

Remark 10.1. In a general bundle setting [Saunders (1989)], the conditions of the above definition are equivalent to saying that the pair $(f, \bar{f})$ is a bundle automorphism of the bundle $\mathbb{R}^k \times Q \to Q$. $\diamond$

Remark 10.2. If we consider the identification between $J^1(Q,\mathbb{R}^k)$ and $J^1\pi_Q$ given in Remark 8.1, the above definition coincides with the definition 4.2.5 in [Saunders (1989)] of the first prolongation of f to the jet bundles.$\diamond$

In a local coordinates system $(x^\alpha, q^i, p^\alpha_i)$ on $J^1(Q,\mathbb{R}^k) \equiv \mathbb{R}^k \times (T^1_k)^*Q$, if $f(x,q) = (f^\alpha(x,q), \bar{f}^i(q))$, then

$$j^{1*}f = \left(f^\alpha(x^\beta, q^j), \bar{f}^i(q^j), \left(\frac{\partial f^\alpha}{\partial q^k} + p^\beta_k \frac{\partial f^\alpha}{\partial x^\beta}\right)\left(\frac{\partial (\bar{f}^{-1})^k}{\partial \bar{q}^i} \circ \bar{f}\right)(q^j)\right), \quad (10.3)$$

where $\bar{q}^i$ are the coordinates on $Q = \bar{f}(Q)$.

Now we are in conditions to give the definition of prolongation of π_Q-projectable vector field.

Definition 10.3. Let $Z \in \mathfrak{X}(\mathbb{R}^k \times Q)$ be a π_Q-projectable vector field, with local 1-parameter group of diffeomorphisms $\sigma_s : \mathbb{R}^k \times Q \to \mathbb{R}^k \times Q$. Then the local 1-parameter group of diffeomorphisms $j^{1*}\sigma_s : \mathbb{R}^k \times (T^1_k)^*Q \to \mathbb{R}^k \times (T^1_k)^*Q$ is generated by a vector field $Z^{1*} \in \mathfrak{X}(\mathbb{R}^k \times (T^1_k)^*Q)$, called ***the natural prolongation (or complete lift) of*** Z ***to*** $\mathbb{R}^k \times (T^1_k)^*Q$.

In local coordinates, if $Z \in \mathfrak{X}(\mathbb{R}^k \times Q)$ is a π_Q-projectable vector field with local expression,

$$Z = Z_\alpha \frac{\partial}{\partial x^\alpha} + Z^i \frac{\partial}{\partial q^i},$$

then from (10.3) and Definition 10.3 we deduce that the natural prolongation Z^{1*} has the following local expression

$$Z^{1*} = Z_\alpha \frac{\partial}{\partial x^\alpha} + Z^i \frac{\partial}{\partial q^i} + \left(\frac{dZ_\alpha}{dq^i} - p^\beta_j \frac{dZ^j}{dq^i}\right)\frac{\partial}{\partial p^\beta_i}, \quad (10.4)$$

where d/dq^i denoted the total derivative, that is,

$$\frac{d}{dq^i} = \frac{\partial}{\partial q^i} + p^\beta_i \frac{\partial}{\partial x^\beta}.$$

Remark 10.3. In the general framework of first order jet bundles, there exists a notion of prolongation of vector field which reduces to Definition 10.3 when one considers the trivial bundle $\pi_Q \colon \mathbb{R}^k \times Q \to Q$. For a full description in the general case, see [Saunders (1989)]. ◇

10.1.2 *Variational principle*

Now we can describe the multiple integral problem from which one obtains the Hamilton-De Donder-Weyl equations.

Definition 10.4. Denote by $Sec_c(\mathbb{R}^k, \mathbb{R}^k \times (T^1_k)^*Q)$ the set of sections with compact support of the bundle

$$\pi_{\mathbb{R}^k} \circ (\pi_Q)_{1,0} : \mathbb{R}^k \times (T^1_k)^*Q \to \mathbb{R}^k.$$

Let $H\colon \mathbb{R}^k \times (T^1_k)^*Q \to \mathbb{R}$ be a Hamiltonian function: then we define the integral action associated to H by

$$\begin{aligned} \mathbb{H} : Sec_c(\mathbb{R}^k, \mathbb{R}^k \times (T^1_k)^*Q) &\to \mathbb{R} \\ \varphi \mapsto \mathbb{H}(\varphi) &= \int_{\mathbb{R}^k} \varphi^*\Theta\,, \end{aligned}$$

where

$$\Theta = \sum_{\alpha=1}^{k} \theta^\alpha \wedge d^{k-1}x_\alpha - H d^k x, \tag{10.5}$$

is a k-form on $\mathbb{R}^k \times (T^1_k)^*Q$ being $d^{k-1}x_\alpha = \iota_{\frac{\partial}{\partial x^\alpha}} d^k x$ and $d^k x = dx^1 \wedge \cdots \wedge dx^k$ as in section 4.1.

Remark 10.4. In the above definition we consider the following commutative diagram

$$\begin{array}{ccc} \mathbb{R}^k \times (T^1_k)^*Q & \xrightarrow[(\pi_Q)_{1,0}]{} & \mathbb{R}^k \times Q \\ \varphi \uparrow & & \downarrow \pi_{\mathbb{R}_k} \\ \mathbb{R}^k & \xrightarrow{Id_{\mathbb{R}^k}} & \mathbb{R}^k \end{array}$$

◇

With the aim to describe the extremals of $\mathbb{H}$ we first prove the following

Lemma 10.1. *Let $\varphi\colon \mathbb{R}^k \to \mathbb{R}^k \times (T^1_k)^*Q$ be an element of $Sec_c(\mathbb{R}^k, \mathbb{R}^k \times (T^1_k)^*Q)$. For each $\pi_{\mathbb{R}^k}$-vertical vector field $Z \in \mathfrak{X}(\mathbb{R}^k \times Q)$ with one-parameter group of diffeomorphism $\{\sigma_s\}$ one has that*

$$\varphi_s := j^{1*}\sigma_s \circ \varphi$$

is a section of the canonical projection $\pi_{\mathbb{R}^k} \circ (\pi_Q)_{1,0} : \mathbb{R}^k \times (T^1_k)^*Q \to \mathbb{R}^k$.

Proof. If Z is $\pi_{\mathbb{R}^k}$-vertical vector field, then one has the following local expression

$$Z(x,q) = Z^i(x,q)\frac{\partial}{\partial q^i}\bigg|_{(x,q)}, \tag{10.6}$$

for some $Z^i \in \mathcal{C}^\infty(\mathbb{R}^k \times Q)$.

Since $\{\sigma_s\}$ is the one-parameter group of diffeomorphism of Z one obtains

$$Z(x,q) = (\sigma_{(x,q)})_*(0)\Big(\frac{d}{ds}\Big|_0\Big)$$
$$= \frac{d(x^\alpha \circ \sigma_{(x,q)})}{ds}\Big|_0 \frac{\partial}{\partial x^\alpha}\Big|_{(x,q)} + \frac{d(q^i \circ \sigma_{(x,q)})}{ds}\Big|_0 \frac{\partial}{\partial q^i}\Big|_{(x,q)} .$$

Comparing (10.6) and the above expression of Z at an arbitrary point $(x,q) \in \mathbb{R}^k \times Q$, we have

$$\frac{d(x^\alpha \circ \sigma_{(x,q)})}{ds}\Big|_0 = 0\,,$$

and then we deduce that

$$(x^\alpha \circ \sigma_{(x,q)})(s) = \text{ constant}\,,$$

but as $\sigma_{(x,q)}(0) = (x,q)$ we know that $(x^\alpha \circ \sigma_{(x,q)})(0) = x^\alpha$ and, thus,

$$(x^\alpha \circ \sigma_{(x,q)})(s) = x^\alpha.$$

Then

$$\sigma_s(x,q) = (x, q^i \circ \sigma_s(x,q)),$$

which implies $\pi_{\mathbb{R}^k} \circ \sigma_s = \pi_{\mathbb{R}^k}$.

Now, from (10.3) one obtains

$$\pi_{\mathbb{R}^k} \circ (\pi_Q)_{1,0} \circ \varphi_s(x) = \pi_{\mathbb{R}^k} \circ (\pi_Q)_{1,0} \circ j^{1\,*}\sigma_s \circ \varphi(x)$$
$$= \pi_{\mathbb{R}^k} \circ (\pi_Q)_{1,0}(x, (\sigma_s)^i_Q(q), p^\alpha_k \frac{\partial((\sigma_s)^{-1}_Q)^k}{\partial q^i} \circ (\sigma_s)_Q) = x$$

that is, φ_s is a section of $\pi_{\mathbb{R}^k} \circ (\pi_Q)_{1,0}$. □

Definition 10.5. A section $\varphi : \mathbb{R}^k \to \mathbb{R}^k \times (T^1_k)^*Q \in Sec_c(\mathbb{R}^k, \mathbb{R}^k \times (T^1_k)^*Q)$, is an **extremal** of $\mathbb{H}$ if

$$\frac{d}{ds}\Big|_{s=0} \mathbb{H}(j^{1*}\sigma_s \circ \varphi) = 0$$

where $\{\sigma_s\}$ is the one-parameter group of diffeomorphism of some $\pi_{\mathbb{R}^k}$-vertical and π_Q-projectable vector field $Z \in \mathfrak{X}(\mathbb{R}^k \times Q)$.

Remark 10.5. In the above definition it is necessary that Z is a $\pi_{\mathbb{R}^k}$-vertical vector field to guarantee that each

$$\varphi_s := j^{1*}\sigma_s \circ \varphi$$

is a section of the canonical projection $\pi_{\mathbb{R}^k} \circ (\pi_Q)_{1,0} : \mathbb{R}^k \times (T^1_k)^*Q \to \mathbb{R}^k$, as we have proved in the above lemma. ◇

The multiple integral variational problem associated to a Hamiltonian H consists to obtain the extremals of the integral action $\mathbb{H}$.

Theorem 10.1. *Let $\varphi \in Sec_c(\mathbb{R}^k, \mathbb{R}^k \times (T^1_k)^*Q)$ and $H\colon \mathbb{R}^k \times T^1_kQ \to \mathbb{R}$ a Hamiltonian function. The following statements are equivalent:*

(1) *φ is an extremal of the variational problem associated to H.*

(2) $\displaystyle\int_{\mathbb{R}^k} \varphi^*\mathcal{L}_{Z^{1*}}\Theta = 0$, *for each vector field $Z \in \mathfrak{X}(\mathbb{R}^k \times Q)$ which is $\pi_{\mathbb{R}^k}$-vertical and π_Q-projectable.*

(3) $\varphi^*\iota_{Z^{1*}}d\Theta = 0$, *for each $\pi_{\mathbb{R}^k}$-vertical and π_Q-projectable vector field Z.*

(4) *If $(U; x^\alpha, q^i, p^\alpha_i)$ is a canonical system of coordinates on $\mathbb{R}^k \times (T^1_k)^*Q$, then φ satisfies the Hamilton-De Donder-Weyl equations (10.1).*

Proof. $(1 \Leftrightarrow 2)$ Let $Z \in \mathfrak{X}(\mathbb{R}^k \times Q)$ be a $\pi_{\mathbb{R}^k}$-vertical and π_Q-projectable vector field. Denote by $\{\sigma_s\}$ the one-parameter group of diffeomorphism associated to Z.

Under these conditions we have

$$\begin{aligned}
\frac{d}{ds}\Big|_{s=0}\mathbb{H}(j^{1*}\sigma_s\circ\varphi) &= \frac{d}{ds}\Big|_{s=0}\int_{\mathbb{R}^k}(j^{1*}\sigma_s\circ\varphi)^*\Theta \\
&= \lim_{s\to 0}\frac{1}{s}\left(\int_{\mathbb{R}^k}(j^{1*}\sigma_s\circ\varphi)^*\Theta - \int_{\mathbb{R}^k}\varphi^*\Theta\right) \\
&= \lim_{s\to 0}\frac{1}{s}\left(\int_{\mathbb{R}^k}\varphi^*\left((j^{1*}\sigma_s)^*\Theta\right) - \int_{\mathbb{R}^k}\varphi^*\Theta\right) \\
&= \lim_{s\to 0}\frac{1}{s}\int_{\mathbb{R}^k}\varphi^*[(j^{1*}\sigma_s)^*\Theta - \Theta] \\
&= \int_{\mathbb{R}^k}\varphi^*\mathcal{L}_{Z^{1*}}\Theta\,.
\end{aligned}$$

Therefore we obtain the equivalence between the items (1) and (2).

$(2 \Leftrightarrow 3)$ Taking into account that

$$\mathcal{L}_{Z^{1*}}\Theta = d\iota_{Z^{1*}}\Theta + \iota_{Z^{1*}}d\Theta\,,$$

one obtains

$$\int_{\mathbb{R}^k}\varphi^*\mathcal{L}_{Z^{1*}}\Theta = \int_{\mathbb{R}^k}\varphi^*d\iota_{Z^{1*}}\Theta + \int_{\mathbb{R}^k}\varphi^*\iota_{Z^{1*}}d\Theta$$

and since φ has compact support, using Stokes's theorem we deduce

$$\int_{\mathbb{R}^k}\varphi^*d\iota_{Z^{1*}}\Theta = \int_{\mathbb{R}^k}d\varphi^*\iota_{Z^{1*}}\Theta = 0\,,$$

and then

$$\int_{\mathbb{R}^k} \varphi^* \mathcal{L}_{Z^{1*}} \Theta = 0$$

(for each Z vector field $\pi_{\mathbb{R}^k}$-vertical) if and only if,

$$\int_{\mathbb{R}^k} \varphi^* \iota_{Z^{1*}} d\Theta = 0\,.$$

But by the fundamental theorem of the variational calculus, the last equality is equivalent to

$$\varphi^* \iota_{Z^{1*}} d\Theta = 0.$$

(3 ⇔ 4) Suppose that

$$\varphi : \mathbb{R}^k \to \mathbb{R}^k \times (T^1_k)^* Q$$

is a section of $\pi_{\mathbb{R}^k} \circ (\pi_Q)_{1,0}$ such that

$$\varphi^* \iota_{Z^{1*}} d\Theta = 0\,,$$

for each $Z \in \mathfrak{X}(\mathbb{R}^k \times Q)$ $\pi_{\mathbb{R}^k}$-vertical and π_Q-projectable vector field.

In canonical coordinates we have

$$Z = Z^i \frac{\partial}{\partial q^i},$$

for some functions $Z^i \in \mathcal{C}^\infty(Q)$ then; from (10.4) one has

$$Z^{1*} = Z^i \frac{\partial}{\partial q^i} - p^\alpha_j \frac{\partial Z^j}{\partial q^i} \frac{\partial}{\partial p^\alpha_i}\,.$$

Therefore,

$$\begin{aligned}
\iota_{Z^{1*}} d\Theta &= \iota_{Z^{1*}} \left(\sum_{\alpha=1}^k dp^\alpha_i \wedge dq^i \wedge d^{k-1}x_\alpha - dH \wedge d^k x \right) \\
&= -Z^i \left(\sum_{\alpha=1}^k dp^\alpha_i \wedge d^{k-1}x_\alpha + \frac{\partial H}{\partial q^i} d^k x \right) \qquad (10.7) \\
&= \sum_{\alpha=1}^k p^\alpha_j \frac{\partial Z^j}{\partial q^i} \left(dq^i \wedge d^{k-1}x_\alpha - \frac{\partial H}{\partial p^\alpha_i} d^k x \right)\,.
\end{aligned}$$

So, if

$$\varphi(x) = (x^\alpha, \psi^i(x), \psi^\alpha_i(x))$$

then $q^i = \psi^i(x)$ and $p_i^\alpha = \psi_i^\alpha(x)$ along the image of φ. Using (10.7) and taking into account that $Z^i(x) := (Z^i \circ \varphi)(x)$ one has,

$$
\begin{aligned}
0 = [\varphi^* \iota_{Z^{1*}} d\Theta](x) &= -Z^i(x) \left(\sum_{\alpha=1}^k \frac{\partial \psi_i^\alpha}{\partial x^\alpha}\Big|_x + \frac{\partial H}{\partial q^i}\Big|_{\varphi(x)} \right) d^k x \\
&- \sum_{\alpha=1}^k \psi_j^\alpha(x) \frac{\partial Z^j}{\partial q^i}\Big|_x \left(\frac{\partial \psi^i}{\partial x^\alpha}\Big|_x - \frac{\partial H}{\partial p_i^\alpha}\Big|_{\varphi(x)} \right) d^k x \\
&= \left[-Z^i(x) \left(\sum_{\alpha=1}^k \frac{\partial \psi_i^\alpha}{\partial x^\alpha}\Big|_x + \frac{\partial H}{\partial q^i}\Big|_{\varphi(x)} \right) \right] d^k x \\
&- \left[\sum_{\alpha=1}^k \psi_j^\alpha(x) \frac{\partial Z^j}{\partial q^i}\Big|_x \left(\frac{\partial \psi^i}{\partial x^\alpha}\Big|_x - \frac{\partial H}{\partial p_i^\alpha}\Big|_{\varphi(x)} \right) \right] d^k x
\end{aligned}
$$

for any $\pi_{\mathbb{R}^k}$-vertical and π_Q-projectable vector field Z.

The above identity is equivalent to the following expression:

$$
\begin{aligned}
& Z^i(x) \left(\sum_{\alpha=1}^k \frac{\partial \psi_i^\alpha}{\partial x^\alpha}\Big|_x + \frac{\partial H}{\partial q^i}\Big|_{\varphi(x)} \right) \\
& + \sum_{\alpha=1}^k \psi_j^\alpha(x) \frac{\partial Z^j}{\partial q^i}\Big|_x \left(\frac{\partial \psi^i}{\partial x^\alpha}\Big|_x - \frac{\partial H}{\partial p_i^\alpha}\Big|_{\varphi(x)} \right) \\
& = 0 \,,
\end{aligned}
$$

for each $Z^i(x^\alpha, q^j)$. Therefore,

$$
\begin{aligned}
Z^i(x) \left(\sum_{\alpha=1}^k \frac{\partial \psi_i^\alpha}{\partial x^\alpha}\Big|_x + \frac{\partial H}{\partial q^i}\Big|_{\varphi(x)} \right) &= 0 \,, \\
\sum_{\alpha=1}^k \psi_j^\alpha(x) \frac{\partial Z^j}{\partial q^i}\Big|_x \left(\frac{\partial \psi^i}{\partial x^\alpha}\Big|_x - \frac{\partial H}{\partial p_i^\alpha}\Big|_{\varphi(x)} \right) &= 0 \,.
\end{aligned}
\tag{10.8}
$$

From the first of the identities of (10.8) one obtains the first set of the Hamilton-De Donder-Weyl field equations, that is,

$$
\sum_{\alpha=1}^k \frac{\partial \psi_i^\alpha}{\partial x^\alpha}\Big|_x = - \frac{\partial H}{\partial q^i}\Big|_{\varphi(x)} \,.
$$

Consider now a coordinate neighborhood $(U; x^\alpha, q^i, p_i^\alpha)$. Since there exists a critical section for each point on U, one obtains that

$$
\frac{\partial Z^j}{\partial q^i}\Big|_x \left(\frac{\partial \psi^i}{\partial x^\alpha}\Big|_x - \frac{\partial H}{\partial p_i^\alpha}\Big|_{\varphi(x)} \right) = 0 \,,
$$

using the second identity of (10.8).

Finally, as the Z^i can be arbitrarily chosen, then $\left.\dfrac{\partial Z^j}{\partial q^i}\right|_x$ can take arbitrary values, and thus we have,

$$\left.\frac{\partial \psi^i}{\partial x^\alpha}\right|_x - \left.\frac{\partial H}{\partial p_i^\alpha}\right|_{\varphi(x)} = 0 \,,$$

which is the second set of the Hamilton-De Donder-Weyl equations.

The converse can be proved by reversing the above arguments. □

Remark 10.6. A. Echeverría-Enríquez *et al.* have obtained in [Echeverría-Enríquez, Muñoz-Lecanda and Román-Roy (2000)] a similar result but considering a variational principle in the multisymplectic setting. ◇

10.2 Hamilton-De Donder-Weyl equations: k-cosymplectic approach

The above variational principle allows us to obtain the HDW equations but as in the case of the Hamiltonian functions independent of the space-time coordinates, there is another method to obtain these equations. In this section we describe these equations which can be obtained using the k-cosymplectic Hamiltonian system when we consider the k-cosymplectic manifold $M = \mathbb{R}^k \times (T_k^1)^*Q$ with the canonical k-cosymplectic structure.

So, we now consider a k-cosymplectic Hamiltonian system

$$(\mathbb{R}^k \times (T_k^1)^*Q, \eta^\alpha, \Omega^\alpha, H),$$

where the Hamiltonian function H is now a function defined on $\mathbb{R}^k \times (T_k^1)^*Q$. From Theorem 9.1 one obtains that if $X = (X_1, \dots, X_k) \in \mathfrak{X}_H^k(\mathbb{R}^k \times (T_k^1)^*Q)$ is an integrable k-vector field and $\varphi\colon U \subset \mathbb{R}^k \to (T_k^1)^*Q$ is an integral section of X, then φ is a solution of the following systems of partial differential equations

$$\left.\frac{\partial H}{\partial q^i}\right|_{\varphi(x)} = -\sum_{\beta=1}^{k} \left.\frac{\partial \psi_i^\beta}{\partial x^\beta}\right|_x, \quad \left.\frac{\partial H}{\partial p_i^\alpha}\right|_{\varphi(x)} = \left.\frac{\partial \psi^i}{\partial x^\alpha}\right|_x,$$

that is φ is a solution of the HDW equations (10.1).

Therefore, the Hamilton-De Donder-Weyl equations are a particular case of the system of partial differential equations which one can obtain from the k-cosymplectic equation.

Remark 10.7. In the case $k = 1$, with $M = \mathbb{R} \times T^*Q$, equation (9.1) reduces to the equations of the non-autonomous Hamiltonian mechanics.

Therefore the formalism described here includes as a particular case the Hamiltonian formalism for non-autonomous mechanics. ◇

Chapter 11

Hamilton-Jacobi Equation

There are several attempts to extend the Hamilton-Jacobi theory for classical field theories. In [de León, Marrero, Martín de Diego, Salgado and Vilariño (2010)] we have described this theory in the framework of the so-called k-symplectic formalism [Awane (1992); Günther (1987); de León, Méndez and Salgado (1988, 1991)]. In this section we consider the k-cosymplectic framework. Another attempts in the framework of the multisymplectic formalism [Cantrijn, Ibort and de León (1999); Kijowski and Tulczyjew (1979)] have been discussed in [de León, Marrero and Martín de Diego (2009); Paufler and Römer (2002,b)].

In classical field theory the Hamilton-Jacobi equation is (see [Rund (1973)])

$$\frac{\partial W^\alpha}{\partial x^\alpha} + H\Big(x^\beta, q^i, \frac{\partial W^\alpha}{\partial q^i}\Big) = 0 \tag{11.1}$$

where $W^1, \dots, W^k \colon \mathbb{R}^k \times Q \to \mathbb{R}$, $1 \leq \alpha \leq k$.

The classical statement of time-dependent Hamilton-Jacobi equation is the following [Abraham and Marsden (1978)]:

Theorem 11.1. *Let $H \colon \mathbb{R} \times T^*Q \to \mathbb{R}$ be a Hamiltonian and T^*Q the symplectic manifold with the canonical symplectic structure $\omega = -d\theta$. Let X_{H_t} be a Hamiltonian vector field on T^*Q associated to the Hamiltonian $H_t \colon T^*Q \to \mathbb{R}$, $H_t(\nu_q) = H(t, \nu_q)$, and $W \colon \mathbb{R} \times Q \to \mathbb{R}$ be a smooth function. The following two conditions are equivalent:*

(1) *for every curve c in Q satisfying*

$$c'(t) = \pi_*\Big(X_{H_t}(dW_t(c(t)))\Big)$$

the curve $t \mapsto dW_t(c(t))$ is an integral curve of X_{H_t}, where $W_t \colon Q \to \mathbb{R}$, $W_t(q) = W(t, q)$.

(2) *W satisfies the Hamilton-Jacobi equation*

$$H(x,q^i,\frac{\partial W}{\partial q^i})+\frac{\partial W}{\partial t} = \text{constant on } T^*Q$$

that is,

$$H_t\circ dW_t+\frac{\partial W}{\partial t}=K(t)\,.$$

Now we will extend this result to classical field theories.

11.1 The Hamilton-Jacobi equation

In this section we introduce a geometric version of the Hamilton-Jacobi theory based on the k-cosymplectic formalism ([de León and Vilariño (2014)]). In the particular case $k=1$ we recover the above theorem for time-dependent classical mechanics.

For each $x=(x^1,\dots,x^k)\in\mathbb{R}^k$ we consider the following mappings

$$\begin{aligned} i_x\colon\ & Q\to\mathbb{R}^k\times Q \\ & q\mapsto (x,q) \end{aligned} \qquad\text{and}\qquad \begin{aligned} j_x\colon\ & (T^1_k)^*Q \to\mathbb{R}^k\times(T^1_k)^*Q \\ & (\nu_{1_q},\dots,\nu_{k_q})\mapsto(x,\nu_{1_q},\dots,\nu_{k_q})\,. \end{aligned}$$

Let $\gamma\colon\mathbb{R}^k\times Q\to\mathbb{R}^k\times(T^1_k)^*Q$ be a section of $(\pi_Q)_{1,0}$. Let us observe that given a section γ is equivalent to giving a section $\bar\gamma\colon\mathbb{R}^k\times Q\to(T^1_k)^*Q$ of $\pi^k\colon(T^1_k)^*Q\to Q$ along the map $\pi_Q\colon\mathbb{R}^k\times Q\to Q$.

If fact, given γ we define $\bar\gamma=\bar\pi_2\circ\gamma$ where $\bar\pi_2$ is the canonical projection $\bar\pi_2\colon\mathbb{R}^k\times(T^1_k)^*Q\to(T^1_k)^*Q$. Conversely, given $\bar\gamma$ we define γ as the composition $\gamma(x,q)=(j_x\circ\bar\gamma)(x,q)$. Now, since $(T^1_k)^*Q$ is the Whitney sum of k copies of the cotangent bundle, giving γ is equivalent to giving a family $(\bar\gamma^1,\dots,\bar\gamma^k)$ of 1-forms along the map π_Q, that is $\pi\circ\gamma^\alpha=\pi_Q$.

If we consider local coordinates $(x^\alpha,q^i,p^\alpha_i)$ we have the following local expressions:

$$\begin{aligned} \gamma(x^\alpha,q^i) &= (x^\alpha,q^i,\gamma^\beta_j(x^\alpha,q^i))\,,\\ \bar\gamma(x^\alpha,q^i) &= (q^i,\gamma^\beta_j(x^\alpha,q^i))\,,\\ \bar\gamma^\alpha(x,q) &= \gamma^\alpha_j(x,q)dq^j(q)\,. \end{aligned} \tag{11.2}$$

Moreover, along this section we suppose that each $\bar\gamma^\alpha$ satisfies that its exterior differential $d\bar\gamma^\alpha$ vanishes over two $\pi_{\mathbb{R}^k}$-vertical vector fields. In local coordinates, using the local expressions (11.2), this condition implies that

$$\frac{\partial\gamma^\alpha_i}{\partial q^j}=\frac{\partial\gamma^\alpha_j}{\partial q^i}\,. \tag{11.3}$$

Now, let $Z = (Z_1, \ldots, Z_k)$ be a k-vector field on $\mathbb{R}^k \times (T^1_k)^*Q$. Using γ we can construct a k-vector field $Z^\gamma = (Z^\gamma_1, \ldots, Z^\gamma_k)$ on $\mathbb{R}^k \times Q$ such that the following diagram is commutative

$$\begin{array}{ccc} \mathbb{R}^k \times (T^1_k)^*Q & \xrightarrow{\quad Z \quad} & T^1_k(\mathbb{R}^k \times (T^1_k)^*Q) \\ \gamma \big\uparrow \big\downarrow (\pi_Q)_{1,0} & & \big\downarrow T^1_k(\pi_Q)_{1,0} \\ \mathbb{R}^k \times Q & \xrightarrow{\quad Z^\gamma \quad} & T^1_k(\mathbb{R}^k \times Q) \end{array}$$

that is,

$$Z^\gamma := T^1_k(\pi_Q)_{1,0} \circ Z \circ \gamma \, .$$

Let us recall that for an arbitrary differentiable map $f : M \to N$, the induced map $T^1_k f : T^1_k M \to T^1_k N$ of f is defined by (5.4).

Let us observe that if Z is integrable then Z^γ is also integrable.

In local coordinates, if each Z_α is locally given by

$$Z_\alpha = (Z_\alpha)_\beta \frac{\partial}{\partial x^\beta} + Z^i_\alpha \frac{\partial}{\partial q^i} + (Z_\alpha)^\beta_i \frac{\partial}{\partial p^\beta_i}$$

then Z^γ_α has the following local expression:

$$Z^\gamma_\alpha = \big((Z_\alpha)_\beta \circ \gamma\big) \frac{\partial}{\partial x^\beta} + (Z^i_\alpha \circ \gamma) \frac{\partial}{\partial q^i} \, . \tag{11.4}$$

In particular, if we consider the k-vector field $R = (R_1, \ldots, R_k)$ given by the Reeb vector fields, we obtain, by a similar procedure, a k-vector field $(R^\gamma_1, \ldots, R^\gamma_k)$ on $\mathbb{R}^k \times Q$. In local coordinates, since

$$R_\alpha = \frac{\partial}{\partial x^\alpha}$$

we have

$$R^\gamma_\alpha = \frac{\partial}{\partial x^\alpha} \, .$$

Next, we consider a Hamiltonian function $H \colon \mathbb{R}^k \times T^1_k Q \to \mathbb{R}$, and the corresponding Hamiltonian system on $\mathbb{R}^k \times T^1_k Q$. Notice that if Z satisfies the Hamilton-De Donder-Weyl equations (10.1), then we have

$$(Z_\alpha)_\beta = \delta_{\alpha\beta} \, ,$$

for all α, β.

Theorem 11.2 (Hamilton-Jacobi theorem). *Let $Z \in \mathfrak{X}^k_H(\mathbb{R}^k \times (T^1_k)^*Q)$ be a k-vector field solution of the k-cosymplectic Hamiltonian equation (9.1) and $\gamma\colon \mathbb{R}^k \times Q \to \mathbb{R}^k \times (T^1_k)^*Q$ be a section of $(\pi_Q)_{1,0}$ satisfying the property described above. If Z is integrable then the following statements are equivalent:*

(1) *If a section $\psi\colon U \subset \mathbb{R}^k \to \mathbb{R}^k \times Q$ of $\pi_{\mathbb{R}^k}\colon \mathbb{R}^k \times Q \to \mathbb{R}^k$ is an integral section of Z^γ, then $\gamma \circ \psi$ is a solution of the Hamilton-De Donder-Weyl equations (10.1);*

(2) $(\pi_Q)^*[d(H \circ \gamma \circ i_x)] + \sum_\alpha \iota_{R^\gamma_\alpha} d\bar{\gamma}^\alpha = 0$ *for all $x \in \mathbb{R}^k$.*

Proof. Let us suppose that a section $\psi\colon U \subset \mathbb{R}^k \to \mathbb{R}^k \times Q$ is an integral section of Z^γ. In local coordinates that means that if $\psi(x) = (x^\alpha, \psi^i(x))$, then

$$[(Z^\gamma_\alpha)^\beta \circ \gamma](\psi(x)) = \delta_\alpha\beta, \quad (Z^i_\alpha \circ \gamma)(\psi(x)) = \frac{\partial \psi^i}{\partial x^\alpha}\,.$$

Now, by hypothesis, $\gamma \circ \psi\colon U \subset \mathbb{R}^k \to \mathbb{R}^k \times (T^1_k)^*Q$ is a solution of the Hamilton-De Donder-Weyl equation for H. In local coordinates, if $\psi(x) = (x, \psi^i(x))$, then $\gamma \circ \psi(x) = (x, \psi^i(x), \gamma^\alpha_i(\psi(x)))$ and, since it is a solution of the Hamilton-De Donder-Weyl equations for H, we have

$$\left.\frac{\partial \psi^i}{\partial x^\alpha}\right|_x = \left.\frac{\partial H}{\partial p^\alpha_i}\right|_{\gamma(\psi(x))} \text{ and } \sum_{\alpha=1}^k \left.\frac{\partial(\gamma^\alpha_i \circ \psi)}{\partial x^\alpha}\right|_x = -\left.\frac{\partial H}{\partial q^i}\right|_{\gamma(\psi(x))}. \tag{11.5}$$

Next, if we compute the differential of the function $H \circ \gamma \circ i_x\colon Q \to \mathbb{R}$, we obtain that:

$$\begin{aligned}&(\pi_Q)^*[d(H \circ \gamma \circ i_x)] + \textstyle\sum_\alpha \iota_{R^\gamma_\alpha} d\bar{\gamma}^\alpha \\ &= \left(\frac{\partial H}{\partial q^i} \circ \gamma \circ i_x + \left(\frac{\partial H}{\partial p^\alpha_j} \circ \gamma \circ i_x\right)\left(\frac{\partial \gamma^\alpha_j}{\partial q^i} \circ i_x\right) + \left(\frac{\partial \gamma^\alpha_i}{\partial x^\alpha} \circ i_x\right)\right) dq^i\,.\end{aligned} \tag{11.6}$$

Therefore, from (11.3), (11.5) and (11.6) and taking into account that one can write $\psi(x) = (i_x \circ \pi_Q \circ \psi)(x)$, where $\pi_Q\colon \mathbb{R}^k \times Q \to Q$ is the canonical projection, we obtain

$$
\begin{aligned}
&((\pi_Q)^*[d(H\circ\gamma\circ i_x)] + \textstyle\sum_\alpha \iota_{R^\gamma_\alpha} d\bar{\gamma}^\alpha)(\pi_Q\circ\psi(x)) \\
&= \left(\frac{\partial H}{\partial q^i}\Big|_{\gamma(\psi(x))} + \frac{\partial H}{\partial p^\alpha_j}\Big|_{\gamma(\psi(x))} \frac{\partial \gamma^\alpha_j}{\partial q^i}\Big|_{\psi(x)} + \frac{\partial \gamma^\alpha_i}{\partial x^\alpha}\Big|_{\psi(x)} \right) dq^i(\pi_Q\circ\psi(x)) \\
&= \left(-\sum_{\alpha=1}^k \frac{\partial(\gamma^\alpha_i\circ\psi)}{\partial x^\alpha}\Big|_x + \frac{\partial \psi^j}{\partial x^\alpha}\Big|_x \frac{\partial \gamma^\alpha_j}{\partial q^i}\Big|_{\psi(x)} + \frac{\partial \gamma^\alpha_i}{\partial x^\alpha}\Big|_{\psi(x)} \right) dq^i(\pi_Q\circ\psi(x)) \\
&= \left(-\sum_{\alpha=1}^k \frac{\partial(\gamma^\alpha_i\circ\psi)}{\partial x^\alpha}\Big|_x + \frac{\partial \psi^j}{\partial x^\alpha}\Big|_x \frac{\partial \gamma^\alpha_i}{\partial q^j}\Big|_{\psi(x)} + \frac{\partial \gamma^\alpha_i}{\partial x^\alpha}\Big|_{\psi(x)} \right) dq^i(\pi_Q\circ\psi(x)) \\
&= 0\,.
\end{aligned}
$$

As we have mentioned above, since Z is integrable, the k-vector field Z^γ is also integrable, and then for each point $(x,q)\in\mathbb{R}^k\times Q$ we have an integral section $\psi\colon U\subset\mathbb{R}^k\to\mathbb{R}^k\times Q$ of Z^γ passing trough that point. Therefore, for any $x\in\mathbb{R}^k$, we get

$$(\pi_Q)^*[d(H\circ\gamma\circ i_x)] + \sum_\alpha \iota_{R^\gamma_\alpha} d\bar{\gamma}^\alpha = 0\,.$$

Conversely, let us suppose that $(\pi_Q)^*[d(H\circ\gamma\circ i_x)] + \sum_\alpha \iota_{R^\gamma_\alpha} d\bar{\gamma}^\alpha = 0$ and take ψ an integral section of Z^γ. We will now prove that $\gamma\circ\psi$ is a solution of the Hamilton-De Donder-Weyl field equations for H.

Since $(\pi_Q)^*[d(H\circ\gamma\circ i_x)] + \sum_\alpha \iota_{R^\gamma_\alpha} d\bar{\gamma}^\alpha = 0$ then from (11.6) we obtain

$$\frac{\partial H}{\partial q^i}\circ\gamma\circ i_x + \left(\frac{\partial H}{\partial p^\alpha_j}\circ\gamma\circ i_x\right)\left(\frac{\partial\gamma^\alpha_j}{\partial q^i}\circ i_x\right) + \left(\frac{\partial \gamma^\alpha_i}{\partial x^\alpha}\circ i_x\right) = 0\,. \qquad (11.7)$$

From (9.2) and (11.4), we know that

$$Z^\gamma_\alpha = \frac{\partial}{\partial x^\alpha} + \left(\frac{\partial H}{\partial p^\alpha_i}\circ\gamma\right)\frac{\partial}{\partial q^i}\,; \qquad (11.8)$$

and then, since $\psi(x,q) = (x,\psi^i(x,q))$ is an integral section of Z^γ, we obtain

$$\frac{\partial \psi^i}{\partial x^\alpha} = \frac{\partial H}{\partial p^\alpha_i}\circ\gamma\circ\psi\,. \qquad (11.9)$$

On the other hand, from (11.3), (11.7) and (11.9) we obtain

$$
\begin{aligned}
&\sum_{\alpha=1}^k \frac{\partial(\gamma^\alpha_i\circ\psi)}{\partial x^\alpha}\Big|_x = \sum_{\alpha=1}^k \left(\frac{\partial\gamma^\alpha_i}{\partial x^\alpha}\Big|_{\psi(x)} + \frac{\partial\gamma^\alpha_i}{\partial q^j}\Big|_{\psi(x)} \frac{\partial\psi^j}{\partial x^\alpha}\Big|_x \right) \\
&= \sum_{\alpha=1}^k \left(\frac{\partial\gamma^\alpha_i}{\partial x^\alpha}\Big|_{\psi(x)} + \frac{\partial\gamma^\alpha_i}{\partial q^j}\Big|_{\psi(x)} \frac{\partial H}{\partial p^\alpha_j}\Big|_{\gamma(\psi(x))} \right) \\
&= \sum_{\alpha=1}^k \left(\frac{\partial\gamma^\alpha_i}{\partial x^\alpha}\Big|_{\psi(x)} + \frac{\partial\gamma^\alpha_j}{\partial q^i}\Big|_{\psi(x)} \frac{\partial H}{\partial p^\alpha_j}\Big|_{\gamma(\psi(x))} \right) = -\frac{\partial H}{\partial q^i}\Big|_{\gamma(\psi(x))}
\end{aligned}
$$

and thus we have proved that $\gamma\circ\psi$ is a solution of the Hamilton-De Donder-Weyl equations. □

Theorem 11.3. *Let $Z \in \mathfrak{X}^k_H(\mathbb{R}^k \times (T^1_k)^*Q)$ be a k-vector field solution of the k-cosymplectic Hamiltonian equation (9.1) and $\gamma\colon \mathbb{R}^k \times Q \to \mathbb{R}^k \times (T^1_k)^*Q$ be a section of $(\pi_Q)_{1,0}$ satisfying the same conditions of the above theorem. Then, the following statements are equivalent:*

(1) $Z|_{Im\,\gamma} - T^1_k\gamma(Z^\gamma) \in \ker\Omega^\sharp \cap \ker\eta^\sharp$, *being $\Omega^\sharp$ and $\eta^\sharp$ the vector bundle morphisms defined in section 10.2.*
(2) $(\pi_Q)^*[d(H\circ\gamma\circ i_x)] + \sum_\alpha \iota_{R^\gamma_\alpha} d\bar{\gamma}^\alpha = 0.$

Proof. A direct computation shows that $Z_\alpha|_{Im\,\gamma} - T\gamma(Z^\gamma_\alpha)$ has the following local expression

$$\left((Z_\alpha)^\beta_j \circ\gamma - \frac{\partial\gamma^\beta_j}{\partial x^\alpha} - (Z^i_\alpha\circ\gamma)\frac{\partial\gamma^\beta_j}{\partial q^i}\right)\frac{\partial}{\partial p^\beta_j}\circ\gamma\,.$$

Thus from (9.6) we know that $Z|_{Im\,\gamma} - T^1_k\gamma(Z^\gamma) \in \ker\Omega^\sharp\cap\ker\eta^\sharp$ if and only if

$$\sum_{\alpha=1}^k\left((Z_\alpha)^\alpha_j\circ\gamma - \frac{\partial\gamma^\alpha_j}{\partial x^\alpha} - (Z^i_\alpha\circ\gamma)\frac{\partial\gamma^\alpha_j}{\partial q^i}\right) = 0\,. \tag{11.10}$$

Now we are ready to prove the result.

Assume that (1) holds, then from (11.3) and (11.10) we obtain

$$\begin{aligned}
0 &= \sum_{\alpha=1}^k\left((Z_\alpha)^\alpha_j\circ\gamma - \frac{\partial\gamma^\alpha_j}{\partial x^\alpha} - (Z^i_\alpha\circ\gamma)\frac{\partial\gamma^\alpha_j}{\partial q^i}\right)\\
&= -\left(\left(\frac{\partial H}{\partial q^j}\circ\gamma\right) + \frac{\partial\gamma^\alpha_j}{\partial x^\alpha} + \left(\frac{\partial H}{\partial p^\alpha_i}\circ\gamma\right)\frac{\partial\gamma^\alpha_j}{\partial q^i}\right)\\
&= -\left(\left(\frac{\partial H}{\partial q^j}\circ\gamma\right) + \frac{\partial\gamma^\alpha_j}{\partial x^\alpha} + \left(\frac{\partial H}{\partial p^\alpha_i}\circ\gamma\right)\frac{\partial\gamma^\alpha_i}{\partial q^j}\right).
\end{aligned}$$

Therefore $(\pi_Q)^*[d(H\circ\gamma\circ i_x)] + \sum_\alpha \iota_{R^\gamma_\alpha} d\bar{\gamma}^\alpha = 0$ (see (11.6)).

The converse is proved in a similar way by reversing the arguments. □

Corollary 11.1. *Let $Z \in \mathfrak{X}^k_H(\mathbb{R}^k \times (T^1_k)^*Q)$ be a solution of (9.1) and $\gamma\colon \mathbb{R}^k\times Q \to \mathbb{R}^k \times (T^1_k)^*Q$ be a section of $(\pi_Q)_{1,0}$ as in the above theorem. If Z is integrable then the following statements are equivalent:*

(1) $Z|_{Im\gamma} - T^1_k\gamma(Z^\gamma) \in \ker\Omega^\sharp\cap\ker\eta^\sharp$;
(2) $(\pi_Q)^*[d(H\circ\gamma\circ i_x)] + \sum_\alpha \iota_{R^\gamma_\alpha} d\bar{\gamma}^\alpha = 0$;

(3) *If a section* $\psi\colon U \subset \mathbb{R}^k \to \mathbb{R}^k \times Q$ *of* $\pi_{\mathbb{R}^k}\colon \mathbb{R}^k \times Q \to \mathbb{R}^k$ *is an integral section of* Z^γ *then* $\gamma \circ \psi$ *is a solution of the Hamilton-De Donder-Weyl equations (10.1).*

Let us observe that there exist k local functions $W^\alpha \in \mathcal{C}^\infty(U)$ such that $\bar{\gamma}^\alpha = dW^\alpha_x$ where the function W^α_x is defined by $W^\alpha_x(q) = W^\alpha(x,q)$. Thus $\gamma^\alpha_i = {}^{\partial W^\alpha}/_{\partial q^i}$ (see [Kobayashi and Nomizu (1963)]). Therefore, the condition

$$(\pi_Q)^*[d(H \circ \gamma \circ i_x)] + \sum_\alpha \iota_{R^\gamma_\alpha} d\bar{\gamma}^\alpha = 0$$

can be equivalently written as

$$\frac{\partial}{\partial q^i}\left(\frac{\partial W^\alpha}{\partial x^\alpha} + H(x^\beta, q^i, \frac{\partial W^\alpha}{\partial q^i})\right) = 0.$$

The above expressions mean that

$$\frac{\partial W^\alpha}{\partial x^\alpha} + H(x^\beta, q^i, \frac{\partial W^\alpha}{\partial q^i}) = K(x^\beta)$$

so that if we put $\tilde{H} = H - K$ we deduce the standard form of the Hamilton-Jacobi equation (since H and $\tilde{H}$ give the same Hamilton-De Donder-Weyl equations)

$$\frac{\partial W^\alpha}{\partial x^\alpha} + \tilde{H}(x^\beta, q^i, \frac{\partial W^\alpha}{\partial q^i}) = 0\,. \tag{11.11}$$

Therefore the equation

$$(\pi_Q)^*[d(H \circ \gamma \circ i_x)] + \sum_\alpha \iota_{R^\gamma_\alpha} d\bar{\gamma}^\alpha = 0 \tag{11.12}$$

can be considered as a geometric version of the ***Hamiton-Jacobi equation for*** k***-cosymplectic field theories.***

11.2 Examples

In this section we shall apply our method to a particular example in classical field theories.

We consider again the equation of a scalar field ϕ (for instance the gravitational field) which acts on the 4-dimensional space-time. Let us recall that its equation is given by (7.42).

We consider the Lagrangian

$$L(x^1,x^2,x^3,x^4,q,v_1,v_2,v_3,v_4) = \sqrt{-g}\Big(F(q) - \frac{1}{2}m^2q^2\Big) + \frac{1}{2}g^{\alpha\beta}v_\alpha v_\beta\,, \tag{11.13}$$

where q denotes the scalar field ϕ and v_α the partial derivative ${}^{\partial\phi}/_{\partial x^\alpha}$. Then equation (7.42) is just the Euler-Lagrange equation associated to L.

Consider the Hamiltonian function $H \in \mathcal{C}^\infty(\mathbb{R}^4 \times (T^1_4)^*\mathbb{R})$ given by

$$H(x^1,x^2,x^3,x^4,q,p^1,p^2,p^3,p^4) = \frac{1}{2\sqrt{-g}} g_{\alpha\beta}p^\alpha p^\beta - \sqrt{-g}\left(F(q) - \frac{1}{2}m^2q^2\right),$$

where (x^1,x^2,x^3,x^4) are the coordinates on $\mathbb{R}^4$, q denotes the scalar field ϕ and $(x^1,x^2,x^3,x^4,q,p^1,p^2,p^3, p^4)$ the canonical coordinates on $\mathbb{R}^4 \times (T^1_4)^*\mathbb{R}$. Let us recall that this Hamiltonian function can be obtained from the Lagrangian L just using the Legendre transformation.

Then

$$\frac{\partial H}{\partial q} = -\sqrt{-g}\Big(F'(q) - m^2q\Big), \quad \frac{\partial H}{\partial p^\alpha} = \frac{1}{\sqrt{-g}} g_{\alpha\beta}p^\beta . \tag{11.14}$$

The Hamilton-Jacobi equation becomes

$$-\sqrt{-g}\Big(F'(q) - m^2q\Big) + \frac{1}{\sqrt{-g}} g_{\alpha\beta}\gamma^\beta \frac{\partial\gamma^\alpha}{\partial q} + \frac{\partial\gamma^\alpha}{\partial x^\alpha} = 0 . \tag{11.15}$$

Since our main goal is to show how the method developed in this chapter works, we will consider, for simplicity, the following particular case:

$$F(q) = \frac{1}{2}m^2q^2,$$

being $(g_{\alpha\beta})$ the Minkowski metric on $\mathbb{R}^4$, i.e., $(g_{\alpha\beta}) = diag(-1,1,1,1)$.

Let $\gamma \colon \mathbb{R}^4 \to \mathbb{R}^4 \times (T^1_k)^*\mathbb{R}$ be the section of $(\pi_\mathbb{R})_{1,0}$ defined by the family of four 1-forms along of $\pi_\mathbb{R} : \mathbb{R}^4 \times \mathbb{R} \to \mathbb{R}$

$$\bar{\gamma}^\alpha = \frac{1}{2}C_\alpha q^2 dq$$

with $1 \le \alpha \le 4$ and where C_α are four constants such that $C_1^2 = C_2^2 + C_3^2 + C_4^2$. This section γ satisfies the Hamilton-Jacobi equation (11.15) that in this particular case is given by

$$-\frac{1}{2}C_1^2q^3 + \frac{1}{2}\sum_{a=2}^{4} C_a^2q^3 = 0 ;$$

therefore, condition (2) of Theorem 11.2 holds.

The 4-vector field $Z^\gamma = (Z_1^\gamma, Z_2^\gamma, Z_3^\gamma, Z_4^\gamma)$ is locally given by

$$Z_1^\gamma = \frac{\partial}{\partial x^1} - \frac{1}{2}C_1q^2\frac{\partial}{\partial q}, \quad Z_a^\gamma = \frac{\partial}{\partial x^a} + \frac{1}{2}C_aq^2\frac{\partial}{\partial q},$$

with $a = 2,3,4$. The map $\psi \colon \mathbb{R}^4 \to \mathbb{R}^4 \times \mathbb{R}$ defined by

$$\psi(x^1,x^2,x^3,x^4) = \frac{2}{C_1x^1 - C_2x^2 - C_3x^3 - C_4x^4 + C}, \quad C \in \mathbb{R},$$

is an integral section of the 4-vector field Z^γ.

By Theorem 11.2 one obtains that the map $\varphi = \gamma \circ \psi$, locally given by

$$(x^\alpha) \to (x^\alpha, \psi(x^\alpha), \frac{1}{2}C_\beta(\psi(x^\alpha))^2),$$

is a solution of the Hamilton-De Donder-Weyl equations associated to H, that is,

$$\begin{aligned} 0 &= \sum_{\alpha=1}^{4} \frac{\partial}{\partial x^\alpha}\left(\frac{1}{2}C_\alpha \psi^2\right), \\ -\tfrac{1}{2}C_1\psi^2 &= \frac{\partial \psi}{\partial x^1}, \\ \tfrac{1}{2}C_a\psi^2 &= \frac{\partial \psi}{\partial x^a}, \quad a = 2,3,4. \end{aligned}$$

Let us observe that these equations imply that the scalar field ψ is a solution to the 3-dimensional wave equation.

In this particular example the functions W^α are given by

$$W^\alpha(x,q) = \frac{1}{6}C_\alpha q^3 + h(x),$$

where $h \in \mathcal{C}^\infty(\mathbb{R}^4)$.

In [Paufler and Römer (2002); Guo and Schmidt (2012)], the authors describe an alternative method that can be compared with the one above.

First, we consider the set of functions $W^\alpha\colon \mathbb{R}^4 \times \mathbb{R} \to \mathbb{R}$, $1 \leq \alpha \leq 4$ defined by

$$W^\alpha(x,q) = (q - \frac{1}{2}\phi(x))\sqrt{-g}g^{\alpha\beta}\frac{\partial \phi}{\partial x^\beta},$$

where ϕ is a solution to the Euler-Lagrange equation (7.42). Using these functions we can consider a section γ of $(\pi_\mathbb{R})_{1,0}\colon \mathbb{R}^4 \times (T^1_4)^*\mathbb{R} \to \mathbb{R}^4 \times \mathbb{R}$ with components

$$\gamma^\alpha = \frac{\partial W^\alpha}{\partial q} = \sqrt{-g}g^{\alpha\beta}\frac{\partial \phi}{\partial x^\beta}.$$

By a direct computation we obtain that this section γ is a solution to the Hamilton-Jacobi equation (11.12).

Now from (11.8) and (11.14) we obtain the 4-vector field Z^γ is given by

$$Z^\gamma_\alpha = \frac{\partial}{\partial x^\alpha} + \frac{\partial \phi}{\partial x^\alpha}\frac{\partial}{\partial q}. \tag{11.16}$$

Let us observe that Z^γ is an integrable 4-vector field on $\mathbb{R}^4 \times \mathbb{R}$. Using the Hamilton-Jacobi theorem we obtain that if $\sigma = (id_{\mathbb{R}^4}, \phi)\colon \mathbb{R}^4 \to \mathbb{R}^4 \times \mathbb{R}$

is an integral section of the 4-vector field Z^γ defined by (11.16), then $\gamma \circ \sigma$ is a solution of the Hamilton-De Donder-Weyl equation associated with the Hamiltonian of the massive scalar field.

If we now consider the particular case $F(q) = m^2 q^2$, we obtain the Klein-Gordon equation; this is just the case discussed in [Paufler and Römer (2002)].

Chapter 12

Lagrangian Classical Field Theories

In a similar way to that developed in chapter 6, we now give a description of the Lagrangian classical field theories using two different approaches: a variational principle and a k-cosymplectic approach.

Given a Lagrangian $L \in \mathcal{C}^\infty(\mathbb{R}^k \times T^1_k Q)$, we shall obtain the local Euler-Lagrange field equations

$$\sum_{\alpha=1}^{k} \frac{\partial}{\partial x^\alpha}\Big|_x \left(\frac{\partial L}{\partial v^i_\alpha}\Big|_{\varphi(x)} \right) = \frac{\partial L}{\partial q^i}\Big|_{\varphi(x)}, \quad v^i_\alpha(\varphi(x)) = \frac{\partial (q^i \circ \varphi)}{\partial x^\alpha}\Big|_x, \tag{12.1}$$

with $\varphi \colon \mathbb{R}^k \to \mathbb{R}^k \times T^1_k Q$. First, we shall use a multiple integral variational problem approach, later we shall give a geometric version of these equations.

Finally, we shall define a Legendre transformation on this new setting which shall allow one to prove the equivalence between both Hamiltonian and Lagrangian formalisms when the Lagrangian satisfies certain regularity condition. We shall use the notation introduced in (7.1) and the notion of prolongations to $\mathbb{R}^k \times T^1_k Q$.

12.1 The stable tangent bundle of k^1-velocities $\mathbb{R}^k \times T^1_k Q$

In chapter 8 we have introduced the model of the so-called k-cosymplectic manifolds that we have used to develop the geometric description of the Hamilton-De Donder-Weyl field equations when the Hamiltonian function depends on the coordinates (x^α) on the base manifold. In this section we introduce its Lagrangian counterpart, i.e., a manifold where we shall develop the k-cosymplectic Lagrangian formalism. Roughly speaking, this manifold is the Cartesian product of the k-dimensional Euclidean space and the tangent bundle of k^1-velocities of an n-dimensional smooth manifold Q.

In this section we shall introduce formally the manifold $\mathbb{R}^k \times T^1_kQ$ and some canonical geometric elements defined on it.

Let us recall that in Remark 6.1 we have introduced the manifold $J^1_0(\mathbb{R}^k, Q)$ of 1-jets of maps from $\mathbb{R}^k$ to Q with source $0 \in \mathbb{R}^k$. In an analogous way fixed a point $x \in \mathbb{R}^k$, we can consider the manifold $J^1_x(\mathbb{R}^k, Q)$ of 1-jets of maps from $\mathbb{R}^k$ to Q with source $x \in \mathbb{R}^k$, i.e.,

$$J^1_x(\mathbb{R}^k, Q) = \bigcup_{q\in Q} J^1_{x,\,q}(\mathbb{R}^k, Q) = \bigcup_{q\in Q} \{j^1_{x,q}\phi \,|\, \phi\colon \mathbb{R}^k \to Q \text{ smooth}, \, \phi(x) = q\}\,.$$

Let $J^1(\mathbb{R}^k, Q)$ be the set of 1-jets from $\mathbb{R}^k$ to Q, that is,

$$J^1(\mathbb{R}^k, Q) = \bigcup_{x\in\mathbb{R}^k} J^1_x(\mathbb{R}^k, Q)\,.$$

This space can be identified with $\mathbb{R}^k \times T^1_kQ$ as follows

$$\begin{array}{ccc} J^1(\mathbb{R}^k, Q) & \to \mathbb{R}^k \times J^1_0(\mathbb{R}^k, Q) \to & \mathbb{R}^k \times T^1_kQ \\ j^1_x\phi = j^1_{x,q}\phi \to & (x, j^1_{0,x}\phi_x) & \to (x, v_{1_q}, \dots, v_{k_q}) \end{array}, \tag{12.2}$$

where $\phi_x(\tilde{x}) = \phi(x + \tilde{x})$, with $\tilde{x} \in \mathbb{R}^k$ and

$$v_{\alpha_q} = (\phi_x)_*(0)\Big(\frac{\partial}{\partial x^\alpha}\Big|_0\Big) = \phi_*(x)\Big(\frac{\partial}{\partial x^\alpha}\Big|_x\Big),$$

being $q = \phi_x(0) = \phi(x)$ and with $1 \le \alpha \le k$.

Therefore, an element in $J^1(\mathbb{R}^k, Q)$ can be thought as a family

$$(x, v_{1_q}, \dots, v_{k_q}) \in \mathbb{R}^k \times T^1_kQ$$

where $x \in \mathbb{R}^k$ and $(v_{1_q}, \dots, v_{k_q}) \in T^1_kQ$. Thus, we can consider the following canonical projections defined by

$$\begin{aligned} &(\pi_{\mathbb{R}^k})_{1,0}(x, v_{1_q}, \dots, v_{k_q}) = (x, q), \; \pi_{\mathbb{R}^k}(x, q) = x, \\ &(\pi_{\mathbb{R}^k})_1(x, v_{1_q}, \dots, v_{k_q}) = x, \qquad \pi_Q(x, q) = q, \\ &p_Q(x, v_{1_q}, \dots, v_{k_q}) = q \end{aligned} \tag{12.3}$$

with $x \in \mathbb{R}^k$, $q \in Q$ and $(v_{1_q}, \dots, v_{k_q}) \in T^1_kQ$. The following diagram illustrates the situation:

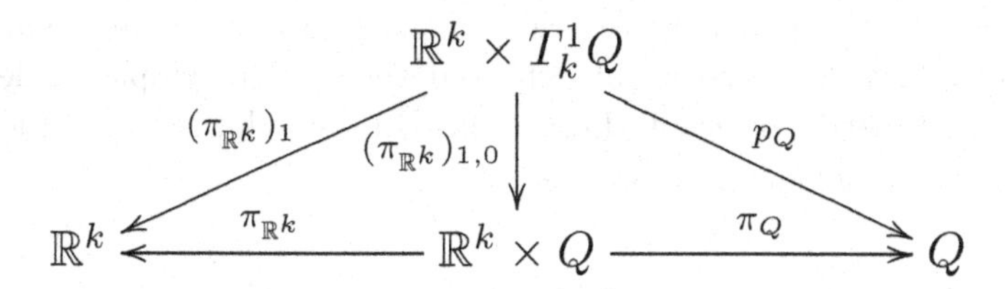

Fig. 12.1 Canonical projections associated to $\mathbb{R}^k \times T^1_k Q$.

If (q^i) with $1 \leq i \leq n$, is a local ***coordinate system*** defined on an open set $U \subset Q$, then the induced local coordinates $(x^\alpha, q^i, v^i_\alpha)$, $1 \leq i \leq n$, $1 \leq \alpha \leq k$ on $\mathbb{R}^k \times T^1_k U = p_Q^{-1}(U)$ are given by

$$\begin{aligned} x^\alpha(x, v_{1_q}, \ldots, v_{k_q}) &= x^\alpha(x) = x^\alpha\,, \\ q^i(x, v_{1_q}, \ldots, v_{k_q}) &= q^i(q)\,, \\ v^i_\alpha(x, v_{1_q}, \ldots, v_{k_q}) &= v_{\alpha_q}(q^i)\,. \end{aligned} \tag{12.4}$$

Thus, $\mathbb{R}^k \times T^1_k Q$ is endowed with a structure of differentiable manifold of dimension $k + n(k+1)$, and the manifold $\mathbb{R}^k \times T^1_k Q$ with the projection $(\pi_{\mathbb{R}^k})_1$ has a structure of vector bundle over $\mathbb{R}^k$.

Considering the identification (12.2), the above coordinates can be defined in terms of 1-jets of maps from $\mathbb{R}^k$ to Q with source in $0 \in \mathbb{R}^k$ as follows

$$\begin{aligned} x^\alpha(j^1_x\phi) &= x^\alpha(x) = x^\alpha \\ q^i(j^1_x\phi) &= q^i(\phi(x)) \\ v^i_\alpha(j^1_x\phi) &= \frac{\partial \phi^i}{\partial x^\alpha}\Big|_x = \phi_*(x)\Big(\frac{\partial}{\partial x^\alpha}\Big|_x\Big)(q^i)\,, \end{aligned}$$

being $\phi\colon \mathbb{R}^k \to Q$.

Remark 12.1. Let us observe that each map $\phi\colon \mathbb{R}^k \to Q$ can be identified with a section $\tilde{\phi}$ of the trivial bundle $\pi_{\mathbb{R}^k} : \mathbb{R}^k \times Q \to \mathbb{R}^k$. Thus the manifold $J^1(\mathbb{R}^k, Q)$ is diffeomorphic to $J^1\pi_{\mathbb{R}^k}$ (see [Saunders (1989)] for a full description of the first-order jet bundle associated to an arbitrary bundle $E \to M$). The diffeomorphism between these two manifolds is given by

$$\begin{aligned} J^1\pi_{\mathbb{R}^k} &\to J^1(\mathbb{R}^k, Q) \\ j^1_x\tilde{\phi} &\to j^1_{x,\phi(x)}\phi \end{aligned}$$

being $\tilde{\phi}\colon \mathbb{R}^k \to \mathbb{R}^k \times Q$ a section of $\pi_{\mathbb{R}^k}$ and $\phi = \pi_Q \circ \tilde{\phi}\colon \mathbb{R}^k \to Q$. ◇

On $\mathbb{R}^k \times T_k^1Q$ there exist several canonical structures which will allow us to introduce the necessary objects to develop a k-cosymplectic description of the Euler-Lagrange field equations. In the following subsections we introduce these geometric elements.

12.1.1 *Canonical tensor fields*

We first define a family $(J^1, \dots, J^k)$ of k tensor fields of type $(1,1)$ on $\mathbb{R}^k \times T_k^1Q$. These tensor fields allow us to define the Poincaré-Cartan forms, in a similar way as in the k-symplectic setting.

To introduce this family we will use the canonical k-tangent structure $\{J^1, \dots, J^k\}$ which we have introduced in section 6.1.1.

For each $1 \leq \alpha \leq k$ we consider the *natural extension* of the tensor fields J^α on T_k^1Q to $\mathbb{R}^k \times T_k^1Q$, (we denote this tensor field also by J^α) whose local expression is

$$J^\alpha = \frac{\partial}{\partial v_\alpha^i} \otimes dq^i. \tag{12.5}$$

Another interesting group of canonical tensors defined on $\mathbb{R}^k \times T_k^1Q$ is the set of canonical vector fields on $\mathbb{R}^k \times T_k^1Q$ defined as follows:

Definition 12.1. The ***Liouville vector field*** Δ on $\mathbb{R}^k \times T_k^1Q$ is the infinitesimal generator of the flow

$$\begin{aligned} \mathbb{R} \times (\mathbb{R}^k \times T_k^1Q) &\longrightarrow \mathbb{R}^k \times T_k^1Q \\ (s, (x, v_{1_q}, \dots, v_{k_q})) &\mapsto (x, e^s v_{1_q}, \dots, e^s v_{k_q}), \end{aligned}$$

and its local expression is

$$\Delta = \sum_{i,A} v_\alpha^i \frac{\partial}{\partial v_\alpha^i}. \tag{12.6}$$

Definition 12.2. The ***canonical vector fields*** $\Delta_1, \dots, \Delta_k$ on $\mathbb{R}^k \times T_k^1Q$ are generator infinitesimals of the flows

$$\begin{aligned} \mathbb{R} \times (\mathbb{R}^k \times T_k^1Q) &\longrightarrow \mathbb{R}^k \times T_k^1Q \\ (s, (x, v_{1_q}, \dots, v_{k_q})) &\mapsto (x, v_{1_q}, \dots, v_{\alpha-1_q}, e^s v_{\alpha_q}, v_{\alpha+1_q}, \dots, v_{k_q}), \end{aligned}$$

for each $\alpha = 1, \dots, k$, respectively. Locally

$$\Delta_\alpha = \sum_{i=1}^n v_\alpha^i \frac{\partial}{\partial v_\alpha^i}, \quad 1 \leq \alpha \leq k. \tag{12.7}$$

From (12.6) and (12.7) we see that $\Delta = \Delta_1 + \ldots + \Delta_k$.

12.1.2 *Prolongation of diffeomorphism and vector fields*

In this section we shall describe how to lift a diffeomorphism of $\mathbb{R}^k \times Q$ to $\mathbb{R}^k \times T^1_kQ$ and, as a consequence, we shall introduce the prolongation of $\pi_{\mathbb{R}^k}$-projectable vector fields on $\mathbb{R}^k \times Q$ to $\mathbb{R}^k \times T^1_kQ$.

Firstly we introduce the following definition of first prolongation of a map $\phi\colon \mathbb{R}^k \to Q$ to $\mathbb{R}^k \times T^1_kQ$.

Definition 12.3. Let $\phi : \mathbb{R}^k \to Q$ be a map, we define the ***first prolongation*** $\phi^{[1]}$ ***of*** ϕ ***to*** $\mathbb{R}^k \times T^1_kQ$ as the map

$$\begin{aligned} \phi^{[1]} : \mathbb{R}^k &\longrightarrow \mathbb{R}^k \times T^1_kQ \\ x &\longrightarrow (x, j^1_0\phi_x) \equiv \left(x, \phi_*(x)\Big(\frac{\partial}{\partial x^1}\Big|_x\Big), \dots, \phi_*(x)\Big(\frac{\partial}{\partial x^k}\Big|_x\Big)\right) \end{aligned}$$

where $\phi_x(y) = \phi(x+y)$.

In local coordinates one has

$$\phi^{[1]}(x^1, \dots, x^k) = (x^1, \dots, x^k, \phi^i(x^1, \dots, x^k), \frac{\partial \phi^i}{\partial x^\alpha}(x^1, \dots, x^k)) . \tag{12.8}$$

Remark 12.2. Let us observe that $\phi^{[1]}$ can be defined as the pair $(id_{\mathbb{R}^k}, \phi^{(1)})$, where $\phi^{(1)}$ is the first prolongation of ϕ to T^1_kQ introduced in Definition 6.7.

Comparing the local expression (12.8) with the second set of the equations (12.1), one observes that a solution $\varphi\colon \mathbb{R}^k \to \mathbb{R}^k \times T^1_kQ$ of the Euler-Lagrange equations (12.1) is of the form $\varphi = \phi^{[1]}$, being ϕ the map given by the composition

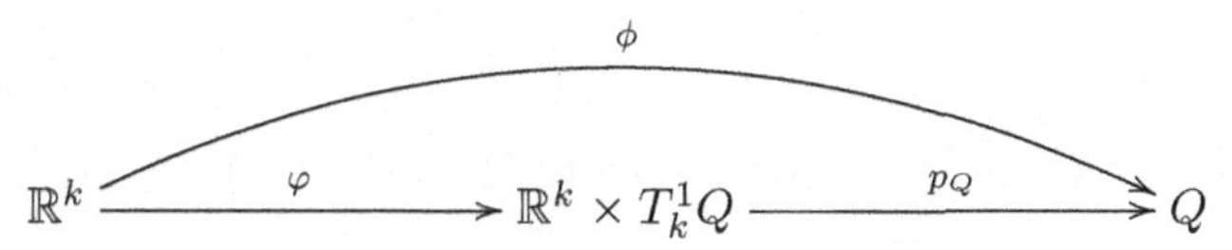

Therefore, equations (12.1) can be written as follows:

$$\sum_{\alpha=1}^k \frac{\partial}{\partial x^\alpha}\left(\frac{\partial L}{\partial x^\alpha}\Big|_{\phi^{[1]}(x)}\right) = \frac{\partial L}{\partial q}\Big|_{\phi^{[1]}(x)}, \tag{12.9}$$

where $1 \leq i \leq n$ and a solution is a map $\phi\colon \mathbb{R}^k \to Q$.

Equations (12.1) are equivalent to

$$\frac{\partial^2 L}{\partial x^\alpha \partial v^i_\alpha}\Big|_{\phi^{[1]}(x)} + \frac{\partial^2 L}{\partial q^j \partial v^i_\alpha}\Big|_{\phi^{[1]}(x)} \frac{\partial \phi^j}{\partial x^\alpha}\Big|_x + \frac{\partial^2 L}{\partial v^j_\beta \partial v^i_\alpha}\Big|_{\phi^{[1]}(x)} \frac{\partial^2 \phi^j}{\partial x^\alpha \partial x^\beta}\Big|_x = \frac{\partial L}{\partial q^i}\Big|_{\phi^{[1]}(x)} .$$

Let us observe that an element in $\mathbb{R}^k \times T^1_kQ$ is of the form $\phi^{[1]}(x)$ for some $\phi\colon \mathbb{R}^k \to Q$ and some $x \in \mathbb{R}^k$. We introduce the prolongation of diffeomorphisms using the first prolongation of maps from $\mathbb{R}^k$ to Q.

Definition 12.4. Let $f\colon \mathbb{R}^k \times Q \to \mathbb{R}^k \times Q$ be a map and $f_{\mathbb{R}^k}\colon \mathbb{R}^k \to \mathbb{R}^k$ be a diffeomorphism, such that $\pi_{\mathbb{R}^k} \circ f = f_{\mathbb{R}^k} \circ \pi_{\mathbb{R}^k}$ [1]. The ***first prolongation*** of f is a map

$$j^1 f\colon J^1(\mathbb{R}^k, Q) \equiv \mathbb{R}^k \times T^1_kQ \to J^1(\mathbb{R}^k, Q) \equiv \mathbb{R}^k \times T^1_kQ$$

defined by

$$(j^1 f)(\phi^{[1]}(x)) = (\pi_Q \circ f \circ \tilde{\phi} \circ f^{-1}_{\mathbb{R}^k})^{[1]}(f_{\mathbb{R}^k}(x)) \tag{12.10}$$

where $\tilde{\phi}$ is the section of $\pi_{\mathbb{R}^k}$ induced by ϕ, that is, $\tilde{\phi} = (id_{\mathbb{R}^k}, \phi)$ and we are considering the first prolongation of the map given by the following composition:

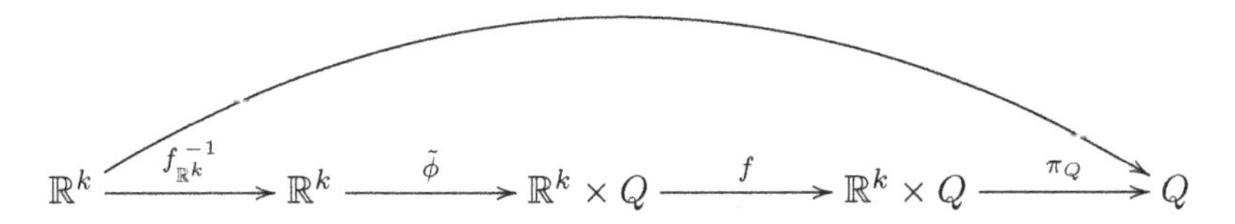

Remark 12.3. If we consider the identification between $J^1(\mathbb{R}^k, Q)$ and $J^1\pi_{\mathbb{R}^k}$ given in Remark 12.1, the above definition coincides with the definition 4.2.5 in [Saunders (1989)] of the first prolongation of f to the jet bundles. ◇

Locally, if $f(x^\alpha, q^i) = (f^\alpha_{\mathbb{R}^k}(x^\beta) = f^\alpha(x^\beta), f^i(x^\beta, q^j))$ then

$$j^1 f(x^\alpha, q^i, v^i_\alpha) = (f^\alpha(x^\beta), f^i(x^\beta, q^j), \frac{df^i}{dx^\beta}\Big(\frac{\partial (f^{-1}_{\mathbb{R}^k})^\beta}{\partial \bar{x}^\alpha} \circ f_{\mathbb{R}^k}(x^\gamma)\Big)\Big), \tag{12.11}$$

where $(\bar{x}^1, \dots, \bar{x}^k)$ are the coordinates on $\mathbb{R}^k = f_{\mathbb{R}^k}(\mathbb{R}^k)$ and df^i/dx^β is the total derivative defined by

$$\frac{df^i}{dx^\beta} = \frac{\partial f^i}{\partial x^\beta} + v^j_\beta \frac{\partial f^i}{\partial q^j}\,.$$

Using the prolongation of diffeomorphism we can define the prolongation of vector field to $\mathbb{R}^k \times T^1_kQ$ in an analogous way that we did in section 10.1.1. Given a $\pi_{\mathbb{R}^k}$-projectable vector field $Z \in \mathfrak{X}(\mathbb{R}^k \times Q)$, we can define the canonical prolongation $Z^1 \in \mathfrak{X}(\mathbb{R}^k \times T^1_kQ)$ using the prolongation of

[1] These conditions are equivalent to saying that the pair $(f, f_{\mathbb{R}^k})$ is a bundle automorphism of the bundle $\mathbb{R}^k \times Q \to \mathbb{R}^k$.

the diffeomorphism of the set $\{\sigma_s\}$, being this set the one-parameter group of diffeomorphism of Z.

Locally if $Z \in \mathfrak{X}(\mathbb{R}^k \times Q)$ is a $\pi_{\mathbb{R}^k}$-projectable vector field with local expression

$$Z = Z^\alpha \frac{\partial}{\partial x^\alpha} + Z^i \frac{\partial}{\partial q^i},$$

then from (12.11) we deduce that the natural prolongation Z^1 has the following local expression

$$Z^1 = Z^\alpha \frac{\partial}{\partial x^\alpha} + Z^i \frac{\partial}{\partial q^i} + \left(\frac{dZ^i}{dx^\alpha} - v^i_\beta \frac{dZ^\beta}{dx^\alpha} \right) \frac{\partial}{\partial v^i_\alpha},$$

where d/dx^α denotes the total derivative, that is

$$\frac{d}{dx^\alpha} = \frac{\partial}{\partial x^\alpha} + v^j_\alpha \frac{\partial}{\partial q^j}.$$

12.1.3 *k-vector fields and* SOPDES

In this section we shall consider again the notion of k-vector field introduced in section 3.1, but in this case, $M = \mathbb{R}^k \times T^1_k Q$. Moreover we describe a particular type of vector fields which are very important in the k-cosymplectic Lagrangian description of the field equations.

We consider $M = \mathbb{R}^k \times T^1_k Q$, with local coordinates $(x^\alpha, q^i, v^i_\alpha)$ on an open set U.

A k-vector field $\mathbf{X}$ on $\mathbb{R}^k \times T^1_k Q$ is a family of k vector fields $(X_1, \ldots, X_k)$ where each $X_\alpha \in \mathfrak{X}(\mathbb{R}^k \times T^1_k Q)$. The local expression of a k-vector field on $\mathbb{R}^k \times T^1_k Q$ is given by $(1 \le \alpha \le k)$

$$X_\alpha = (X_\alpha)_\beta \frac{\partial}{\partial x^\beta} + (X_\alpha)^i \frac{\partial}{\partial q^i} + (X_\alpha)^i_\beta \frac{\partial}{\partial v^i_\beta}. \tag{12.12}$$

Let

$$\varphi\colon U_0 \subset \mathbb{R}^k \to \mathbb{R}^k \times T^1_k Q$$

be an integral section of $(X_1, \ldots, X_k)$ with components

$$\varphi(x) = (\psi_\alpha(x), \psi^i(x), \psi^i_\alpha(x)).$$

Then, since

$$\varphi_*(x)\Big(\frac{\partial}{\partial x^\alpha}\Big|_x\Big) = \frac{\partial \psi^\beta}{\partial x^\alpha}\Big|_x \frac{\partial}{\partial x^\beta}\Big|_{\varphi(x)} + \frac{\partial \psi^i}{\partial x^\alpha}\Big|_x \frac{\partial}{\partial q^i}\Big|_{\varphi(x)} + \frac{\partial \psi^i_\beta}{\partial x^\alpha}\Big|_x \frac{\partial}{\partial v^i_\beta}\Big|_{\varphi(x)}$$

the condition (3.2) is locally equivalent to the following system of partial differential equations (condition (3.3))

$$\frac{\partial \psi_\beta}{\partial x^\alpha}\Big|_x = (X_\alpha)_\beta(\varphi(x))\,, \quad \frac{\partial \psi^i}{\partial x^\alpha}\Big|_x = (X_\alpha)^i(\varphi(x))\,, \quad \frac{\partial \psi^i_\beta}{\partial x^\alpha}\Big|_x = (X_\alpha)^i_\beta(\varphi(x)), \tag{12.13}$$

with $1 \leq i \leq n$ and $1 \leq \alpha, \beta \leq k$. Next, we shall characterize the integrable k-vector fields on $\mathbb{R}^k \times T^1_k Q$ whose integral sections are canonical prolongations of maps from $\mathbb{R}^k$ to Q.

Definition 12.5. A k-vector field $\mathbf{X} = (X_1, \dots, X_k)$ on $\mathbb{R}^k \times T^1_k Q$ is a ***second order partial differential equation*** (SOPDE for short) if

$$\eta^\alpha(X_\beta) = \delta^\alpha_\beta$$

and

$$J^\alpha(X_\alpha) = \Delta_\alpha\,,$$

for all $1 \leq \alpha, \beta \leq k$.

Let (q^i) be a coordinate system on Q and $(x^\alpha, q^i, v^i_\alpha)$ the induced coordinate system on $\mathbb{R}^k \times T^1_k Q$. From (12.5) and (12.7) we deduce that the local expression of a SOPDE $(X_1, \dots, X_k)$ is

$$X_\alpha = \frac{\partial}{\partial x^\alpha} + v^i_\alpha \frac{\partial}{\partial q^i} + (X_\alpha)^i_\beta \frac{\partial}{\partial v^i_\beta}\,, \tag{12.14}$$

where $(X_\alpha)^i_\beta$ are functions on $\mathbb{R}^k \times T^1_k Q$. As a direct consequence of the above local expressions, we deduce that the family of vector fields $\{X_1, \dots, X_k\}$ are linearly independent.

Lemma 12.1. *Let $(X_1, \dots, X_k)$ be a* SOPDE. *A map $\varphi : \mathbb{R}^k \to \mathbb{R}^k \times T^1_k Q$, given by*

$$\varphi(x) = (\psi_\alpha(x), \psi^i(x), \psi^i_\alpha(x))$$

is an integral section of $(X_1, \dots, X_k)$ if, and only if,

$$\psi_\alpha(x) = x^\alpha + c^\alpha\,, \quad \psi^i_\alpha(x) = \frac{\partial \psi^i}{\partial x^\alpha}\Big|_x\,, \quad \frac{\partial^2 \psi^i}{\partial x^\alpha \partial x^\beta}\Big|_x = (X_\alpha)^i_\beta(\varphi(x))\,, \tag{12.15}$$

where c^α is a constant.

Proof. Equations (12.15) follow directly from (12.13) and (12.14). □

Remark 12.4. The integral sections of a SOPDE are given by

$$\varphi(x) = \left(x^\alpha + c^\alpha, \psi^i(x), \frac{\partial \psi^i}{\partial x^\alpha}(x)\right),$$

where the functions $(\psi^i(x))$ satisfy the equation

$$\left.\frac{\partial^2\psi^i}{\partial x^\alpha \partial x^\beta}\right|_x = (X_\alpha)^i_\beta(\psi(x))$$

in (12.15), and the c^α's are constants.

In the particular case when $c = 0$, we have that $\varphi = \phi^{[1]}$ where

$$\phi = p_Q \circ \varphi : \mathbb{R}^k \overset{\varphi}{\to} \mathbb{R}^k \times T^1_k Q \overset{p_Q}{\to} Q$$

that is, $\phi(x) = (\psi^i(x))$. ◇

Lemma 12.2. *Let $\mathbf{X} = (X_1, \dots, X_k)$ be an integrable k-vector field on $\mathbb{R}^k \times T^1_k Q$. If every integral section of $\mathbf{X}$ is the first prolongation $\phi^{[1]}$ of map $\phi : \mathbb{R}^k \to Q$, then $\mathbf{X}$ is a* SOPDE.

Proof. Let us suppose that each X_α is locally given by

$$X_\alpha = (X_\alpha)_\beta \frac{\partial}{\partial x^\beta} + (X_\alpha)^i \frac{\partial}{\partial q^i} + (X_\alpha)^i_\beta \frac{\partial}{\partial v^i_\beta} \,. \tag{12.16}$$

Let $\psi = \phi^{[1]} : \mathbb{R}^k \to \mathbb{R}^k \times T^1_k Q$ be an integral section of $\mathbf{X}$, then from (12.3), (12.13), (12.14) and (12.16), we obtain

$$(X_\alpha)_\beta(\phi^{[1]}(x)) = \delta^\beta_\alpha, \quad (X_\alpha)^i(\phi^{[1]}(x)) = \left.\frac{\partial \phi^i}{\partial x^\alpha}\right|_x = v^i_\alpha(\phi^{[1]}(x))$$

and

$$(X_\alpha)^i_\beta(\phi^{[1]}(x)) = \left.\frac{\partial^2\phi^i}{\partial x^\alpha \partial x^\beta}\right|_x$$

thus X_α is locally given as in (12.14) and then it is a SOPDE. □

12.2 Variational principle

In this section we describe the problem in the setting of the calculus of variations for multiple integrals, which allows us to obtain the Euler-Lagrange field equations. The procedure is similar to section 6.2 but in this case the Lagrangian function depends on the coordinates of the basis space, that is, L is defined on $\mathbb{R}^k \times T^1_k Q$. In particular, if L does not depend on the space-time coordinates we obtain again the results in section 6.2.

Let us observe that given a section ϕ of $\pi_{\mathbb{R}^k} : U_0 \subset \mathbb{R}^k \times Q \to \mathbb{R}^k$, it can be identified with the pair

$$\bar{\phi} = (id_{\mathbb{R}^k}, \phi)$$

where $\bar{\phi} = \pi_Q \circ \phi$. Therefore, any section $\bar{\phi}$ of $\pi_{\mathbb{R}^k}$ can be identified with a map $\phi\colon \mathbb{R}^k \to Q$. Along this section we consider this identification.

Definition 12.6. Let $L\colon \mathbb{R}^k \times T^1_k Q \to \mathbb{R}$ be a Lagrangian. Denote by $Sec_C(\mathbb{R}^k, \mathbb{R}^k \times Q)$ the set of sections of

$$\pi_{\mathbb{R}^k} : U_0 \subset \mathbb{R}^k \times Q \to \mathbb{R}^k$$

with compact support. We define the action associated to L by:

$$\begin{aligned} \mathbb{S} : Sec_C(\mathbb{R}^k, \mathbb{R}^k \times Q) &\to \mathbb{R} \\ \bar{\phi} \qquad\qquad &\mapsto \mathbb{S}(\bar{\phi}) = \int_{\mathbb{R}^k} (\phi^{[1]})^*(L d^k x)\,. \end{aligned}$$

Lemma 12.3. *Let $\bar{\phi} \in Sec_C(\mathbb{R}^k, \mathbb{R}^k \times Q)$ be a section with compact support. If $Z \in \mathfrak{X}(\mathbb{R}^k \times Q)$ is $\pi_{\mathbb{R}^k}$-vertical then*

$$\bar{\phi}_s\colon = \tau_s \circ \bar{\phi}$$

is a section of $\pi_{\mathbb{R}^k} : \mathbb{R}^k \times Q \to \mathbb{R}^k$.

Proof. Since $Z \in \mathfrak{X}(\mathbb{R}^k \times Q)$ is $\pi_{\mathbb{R}^k}$-vertical, then it has the following local expression

$$Z(x,q) = Z^i(x,q) \frac{\partial}{\partial q^i}\Big|_{(x,q)} . \tag{12.17}$$

Now, if $\{\tau_s\}$ is the one-parameter group of diffeomorphisms generated by Z, then, one has

$$\begin{aligned} Z(x,q) &= (\tau_{(x,q)})_*(0)\Big(\frac{d}{ds}\Big|_0\Big) \\ &= \frac{d(x^\alpha \circ \tau_{(x,q)})}{ds}\Big|_0 \frac{\partial}{\partial x^\alpha}\Big|_{(x,q)} + \frac{d(q^i \circ \tau_{(x,q)})}{ds}\Big|_0 \frac{\partial}{\partial q^i}\Big|_{(x,q)} . \end{aligned}$$

Comparing (12.17) and the above expression of Z, and taking into account that it is valid for any point $(x,q) \in \mathbb{R}^k \times Q$, one has

$$\frac{d(x^\alpha \circ \tau_{(x,q)})}{ds} = 0\,.$$

Then

$$(x^\alpha \circ \tau_{(x,q)})(s) = \text{constant}.$$

Moreover $\tau_{(x,q)}(0) = (x,q)$, and we obtain that

$$(x^\alpha \circ \tau_{(x,q)})(0) = x^\alpha\,.$$

Thus $(x^\alpha \circ \tau_{(x,q)})(s) = x^\alpha$ or $(x^\alpha \circ \tau_s)(x,q) = x^\alpha$, and hence

$$\pi_{\mathbb{R}^k} \circ \tau_s = \pi_{\mathbb{R}^k} .$$

Therefore, taking into account this identity we deduce that $\bar{\phi}_s$ is a section of $\pi_{\mathbb{R}^k}$. In fact,

$$\pi_{\mathbb{R}^k} \circ \bar{\phi}_s = \pi_{\mathbb{R}^k} \circ \tau_s \circ \bar{\phi} = \pi_{\mathbb{R}^k} \circ \bar{\phi} = id_{\mathbb{R}^k} ,$$

where in the last identity we use that $\bar{\phi}$ is a section of $\pi_{\mathbb{R}^k}$. □

Definition 12.7. A section $\bar{\phi} = (id_{\mathbb{R}^k}, \phi)\colon \mathbb{R}^k \to \mathbb{R}^k \times Q$, such that $\bar{\phi} \in Sec_C(\mathbb{R}^k, \mathbb{R}^k \times Q)$, is an **extremal** of $\mathbb{S}$ if

$$\frac{d}{ds}\Big|_{s=0} \mathbb{S}(\tau_s \circ \bar{\phi}) = 0$$

where $\{\tau_s\}$ is the one-parameter group of diffeomorphism for some $\pi_{\mathbb{R}^k}$-vertical vector field $Z \in \mathfrak{X}(\mathbb{R}^k \times Q)$.

The variational problem associated with L consists in calculate the extremals of the action $\mathbb{S}$.

Theorem 12.1. *Let* $\bar{\phi} = (id_{\mathbb{R}^k}, \phi) \in Sec_C(\mathbb{R}^k, \mathbb{R}^k \times Q)$ *and* $L : \mathbb{R}^k \times T^1_k Q \to \mathbb{R}$ *be a Lagrangian. The following statements are equivalent:*

(1) $\bar{\phi}$ *is an extremal of* $\mathbb{S}$.

(2) $\displaystyle\int_{\mathbb{R}^k} (\phi^{[1]})^*(\mathcal{L}_{Z^1}(Ld^k x)) = 0$, *for each* $\pi_{\mathbb{R}^k}$*-vertical* $Z \in \mathfrak{X}(\mathbb{R}^k \times Q)$.

(3) ϕ *is a solution of the Euler-Lagrange field equations (12.9).*

Proof. $(1 \Leftrightarrow 2)$ Let $Z \in \mathfrak{X}(\mathbb{R}^k \times Q)$ a $\pi_{\mathbb{R}^k}$-vertical vector field and $\{\tau_s\}$ the one-parameter group of diffeomorphism associated to Z.

Along this proof we denote by ϕ_s the composition $\phi_s = \pi_Q \circ \tau_s \circ \bar{\phi}$. Let us observe that $\bar{\phi}_s = \tau_s \circ \bar{\phi} = (id_{\mathbb{R}^k}, \phi_s)$.

Since $\phi_s^{[1]} = (\pi_Q \circ \tau_s \circ \bar{\phi})_Q^{[1]} = j^1\tau_s \circ \phi^{[1]}$, then

$$\begin{aligned}
\frac{d}{ds}\Big|_{s=0} \mathbb{S}(\tau_s \circ \phi) &= \frac{d}{ds}\Big|_{s=0} \int_{\mathbb{R}^k} ((\phi_s)^{[1]})^*(Ld^k x) \\
&= \lim_{s\to 0} \frac{1}{h}\left(\int_{\mathbb{R}^k} ((\phi_s)^{[1]})^*(Ld^k x) - \int_{\mathbb{R}^k} ((\phi_0)^{[1]})^*(Ld^k x)\right) \\
&= \lim_{s\to 0} \frac{1}{h}\left(\int_{\mathbb{R}^k} ((\pi_Q \circ \tau_s \circ \phi)^{[1]})^*(Ld^k x) - \int_{\mathbb{R}^k} (\phi^{[1]})^*(Ld^k x)\right)
\end{aligned}$$

$$= \lim_{s\to 0} \frac{1}{h} \left(\int_{\mathbb{R}^k} (\phi^{[1]})^* (j^1\tau_s)^* (Ld^k x) - \int_{\mathbb{R}^k} (\phi^{[1]})^* (Ld^k x) \right)$$
$$= \lim_{s\to 0} \frac{1}{h} \int_{\mathbb{R}^k} (\phi^{[1]})^* [(j^1\tau_s)^*(Ld^k x) - (Ld^k x)]$$
$$= \int_{\mathbb{R}^k} (\phi^{[1]})^* \mathcal{L}_{Z^1}(Ld^k x)\,,$$

which implies the equivalence between items (1) and (2).

(2 $\Leftrightarrow$ 3) We know that ϕ is an extremal or critical section of $\mathbb{S}$ if and only if for each $\pi_{\mathbb{R}^k}$-vertical vector field Z one has

$$\int_{\mathbb{R}^k} (\phi^{[1]})^* (\mathcal{L}_{Z^1}(Ld^k x)) = 0\,.$$

Taking into account the identity

$$\mathcal{L}_{Z^1}(Ld^k x) = \iota_{Z^1}(dL \wedge d^k x) + d\iota_{Z^1}(Ld^k x) \tag{12.18}$$

and since ϕ has compact support, from Stokes' theorem one deduces that

$$\int_{\mathbb{R}^k} (\phi^{[1]})^* d\iota_{Z^1}(Ld^k x) = \int_{\mathbb{R}^k} d\big((\phi^{[1]})^* \iota_{Z^1}(Ld^k x)\big) = 0\,. \tag{12.19}$$

Thus, from (12.18) and (12.19) we obtain that $\bar{\phi}$ is an extremal if and only if

$$\int_{\mathbb{R}^k} (\phi^{[1]})^* \iota_{Z^1}(dL \wedge d^k x) = 0\,.$$

If $Z_{(x,q)} = Z^i(x,q) \dfrac{\partial}{\partial q^i}\Big|_{(x,q)}$ then the local expression of Z^1 is

$$Z^1 = Z^i \frac{\partial}{\partial q^i} + \left(\frac{\partial Z^i}{\partial x^\alpha} + \frac{\partial Z^i}{\partial q^j} v^j_\alpha \right) \frac{\partial}{\partial v^i_\alpha}\,,$$

therefore

$$\iota_{Z^1}(dL \wedge d^k x) = \left(Z^i \frac{\partial L}{\partial q^i} + \left(\frac{\partial Z^i}{\partial x^\alpha} + \frac{\partial Z^i}{\partial q^j} v^j_\alpha \right) \frac{\partial L}{\partial v^i_\alpha} \right) d^k x\,. \tag{12.20}$$

Then, from (12.20) we obtain

$$[(\phi^{[1]})^* \iota_{Z^1}(dL \wedge d^k x)](q)$$
$$= \left((Z^i \circ \bar{\phi})(x) \frac{\partial L}{\partial q^i}\Big|_{\phi^{[1]}(x)} + \left(\frac{\partial Z^i}{\partial x^\alpha}\Big|_{\bar{\phi}(x)} + \frac{\partial Z^i}{\partial q^j}\Big|_{\bar{\phi}(x)} \frac{\partial \phi^j}{\partial x^\alpha}\Big|_x \right) \frac{\partial L}{\partial v^i_\alpha}\Big|_{\phi^{[1]}(x)} \right) d^k x\,. \tag{12.21}$$

Let us observe that the last term of (12.21) satisfies

$$\frac{\partial Z^i}{\partial q^j}\Big|_{\bar\phi(x)}\frac{\partial \phi^j}{\partial x^\alpha}\Big|_x\frac{\partial L}{\partial v^i_\alpha}\Big|_{\phi^{[1]}(x)}d^kx = \left(\frac{\partial (Z^i\circ\bar\phi)}{\partial x^\alpha}\Big|_x - \frac{\partial Z^i}{\partial x^\alpha}\Big|_{\bar\phi(x)}\right)\frac{\partial L}{\partial v^i_\alpha}\Big|_{\phi^{[1]}(x)}d^kx\,.$$

After an easy computation we obtain

$$\begin{aligned}&\int_{\mathbb{R}^k}(\phi^{[1]})^*\iota_{Z^1}(dL\wedge d^kx)\\ &= \int_{\mathbb{R}^k}(Z^i\circ\phi)\frac{\partial L}{\partial q^i}\Big|_{\phi^{[1]}(x)}d^kx + \int_{\mathbb{R}^k}\frac{\partial (Z^i\circ\phi)}{\partial x^\alpha}\Big|_x\frac{\partial L}{\partial v^i_\alpha}\Big|_{\phi^{[1]}(x)}d^kx\,.\end{aligned}$$

Since $\bar\phi$ has compact support and using integration by parts, we have

$$\int_{\mathbb{R}^k}\frac{\partial (Z^i\circ\phi)}{\partial x^\alpha}\Big|_x\frac{\partial L}{\partial v^i_\alpha}\Big|_{\phi^{[1]}(x)}d^kx = -\int_{\mathbb{R}^k}(Z^i\circ\phi)(x)\frac{\partial}{\partial x^\alpha}\left(\frac{\partial L}{\partial v^i_\alpha}\Big|_{\phi^{[1]}(x)}\right)d^kx$$

and thus,

$$\begin{aligned}&\int_{\mathbb{R}^k}(\phi^{[1]})^*\iota_{Z^1}(dL\wedge d^kx)\\ &= \int_{\mathbb{R}^k}(Z^i\circ\phi)(x)\left(\frac{\partial L}{\partial q^i}\Big|_{\phi^{[1]}(x)} - \frac{\partial}{\partial x^\alpha}\left(\frac{\partial L}{\partial v^i_\alpha}\Big|_{\phi^{[1]}(x)}\right)\right)d^kx = 0\,.\end{aligned}$$

Since the functions Z^i are arbitrary, from the last identity we obtain the Euler-Lagrange field equations,

$$\frac{\partial L}{\partial q^i}\Big|_{\phi^{[1]}(x)} - \frac{\partial}{\partial x^\alpha}\left(\frac{\partial L}{\partial v^i_\alpha}\Big|_{\phi^{[1]}(x)}\right) = 0\,,\quad 1\leq i\leq n\,.$$

Remark 12.5. In [Echeverría-Enríquez, Muñoz-Lecanda and Román-Roy (1996)] the authors have considered a more general situation; instead of $\mathbb{R}^k\times T^1_kQ\to\mathbb{R}^k$, they consider an arbitrary fiber-bundle $E\to M$. ◇

12.3 *k*-cosymplectic version of Euler-Lagrange field equations

In this section we give the k-cosymplectic description of the Euler-Lagrange field equations (12.1). With this purpose, we introduce some geometric elements associated to a Lagrangian function $L\colon\mathbb{R}^k\times T^1_kQ\to\mathbb{R}$.

12.3.1 *Poincaré-Cartan forms on* $\mathbb{R}^k \times T^1_k Q$

In a similar way that in the k-symplectic approach, one can define a family of 1-forms $\Theta^1_L, \dots, \Theta^k_L$ on $\mathbb{R}^k \times T^1_k Q$ associated with the Lagrangian function $L : \mathbb{R}^k \times T^1_k Q \to \mathbb{R}$, using the canonical tensor fields $J^1, \dots, J^k$ defined in (12.5). Indeed, we put

$$\Theta^\alpha_L = dL \circ J^\alpha \,, \tag{12.22}$$

with $1 \leq \alpha \leq k$. The exterior differential of these 1-forms allows us to consider the family of 2-forms on $\mathbb{R}^k \times T^1_k Q$ by

$$\Omega^\alpha_L = -\, d\Theta^\alpha_L \,. \tag{12.23}$$

From (12.5) and (12.22) we obtain that Θ^α_L is locally given by

$$\Theta^\alpha_L = \frac{\partial L}{\partial v^i_\alpha}\, dq^i \,, \quad 1 \leq \alpha \leq k \tag{12.24}$$

and from (12.23) and (12.24) we obtain that Ω^α_L is locally given by

$$\Omega^\alpha_L = \frac{\partial^2 L}{\partial x^\beta \partial v^i_\alpha}\, dq^i \wedge dx^\beta + \frac{\partial^2 L}{\partial q^j \partial v^i_\alpha}\, dq^i \wedge dq^j + \frac{\partial^2 L}{\partial v^i_\beta \partial v^i_\alpha}\, dq^i \wedge dv^i_\beta \;. \tag{12.25}$$

An important case is when the Lagrangian is regular, i.e., when

$$\det \left(\frac{\partial^2 L}{\partial v^i_\alpha \partial v^j_\beta} \right) \neq 0 \,.$$

The following proposition gives a characterization of the regular Lagrangians.

Prop 12.1. [de León, Merino and Salgado (2001)] Given a Lagrangian function on $\mathbb{R}^k \times T^1_k Q$, the following conditions are equivalent:

(1) L is regular.

(2) $(dx^\alpha, \Omega^1_L, \dots, \Omega^k_L, V)$ is a k-cosymplectic structure on $\mathbb{R}^k \times T^1_k Q$, where

$$V = \ker((\pi_{\mathbb{R}^k})_{1,0})_* = span \left\{ \frac{\partial}{\partial v^i_1}, \dots, \frac{\partial}{\partial v^i_k} \right\}$$

with $1 \leq i \leq n$, is the vertical distribution of the vector bundle $(\pi_{\mathbb{R}^k})_{1,0} : \mathbb{R}^k \times T^1_k Q \to \mathbb{R}^k \times Q$.

12.3.2 *k-cosymplectic Euler-Lagrange equation*

We recall the k-cosymplectic formulation of the Euler-Lagrange equations (12.9) developed by M. de León *et al.* in [de León, Merino and Salgado (2001)].

Let us consider the equations

$$
\begin{aligned}
& dx^\alpha(X_\beta) = \delta^\alpha_\beta, \quad 1 \leq \alpha, \beta \leq k\,, \\
& \sum_{\alpha=1}^{k} \iota_{X_\alpha}\Omega^\alpha_L = dE_L + \sum_{\alpha=1}^{k} \frac{\partial L}{\partial x^\alpha} dx^\alpha
\end{aligned}
\tag{12.26}
$$

where $E_L = \Delta(L) - L$ and denote by $\mathfrak{X}^k_L(\mathbb{R}^k \times T^1_k Q)$ the set of k-vector fields $\mathbf{X} = (X_1, \ldots, X_k)$ on $\mathbb{R}^k \times T^1_k Q$ that are solutions of (12.26).

Let us suppose that $(X_1, \ldots X_k) \in \mathfrak{X}^k_L(\mathbb{R}^k \times T^1_k Q)$ and that each X_α is locally given by

$$
X_\alpha = (X_\alpha)_\beta \frac{\partial}{\partial x^\beta} + (X_\alpha)^i \frac{\partial}{\partial q^i} + (X_\alpha)^i_\beta \frac{\partial}{\partial v^i_\beta}, \quad 1 \leq \alpha \leq k\,.
$$

Equations (12.26) are locally equivalent to the equations

$$
\begin{aligned}
(X_\alpha)_\beta &= \delta^\beta_\alpha\,, \\
(X_\beta)^i \, \frac{\partial^2 L}{\partial x^\alpha \partial v^i_\beta} &= v^i_\beta \, \frac{\partial^2 L}{\partial x^\alpha \partial v^i_\beta}\,, \\
(X_\gamma)^j \, \frac{\partial^2 L}{\partial v^i_\beta \partial v^j_\gamma} &= v^j_\gamma \, \frac{\partial^2 L}{\partial v^i_\beta \partial v^j_\gamma}\,,
\end{aligned}
\tag{12.27}
$$

$$
\begin{aligned}
\frac{\partial^2 L}{\partial q^j \partial v^i_\beta} \left(v^j_\beta - (X_\beta)^j \right) &+ \frac{\partial^2 L}{\partial x^\beta \partial v^i_\beta} \\
&+ v^k_\beta \frac{\partial^2 L}{\partial q^k \partial v^i_\beta} + (X_\beta)^k_\gamma \frac{\partial^2 L}{\partial v^k_\gamma \partial v^i_\beta} = \frac{\partial L}{\partial q^i}\,.
\end{aligned}
$$

If L is regular then these equations are transformed into the following

$$
(X_\alpha)_\beta = \delta^\beta_\alpha, \quad (X_\alpha)^i = v^i_\alpha, \quad \sum_{\alpha=1}^{k} X_\alpha\Big(\frac{\partial L}{\partial v^i_\alpha}\Big) = \frac{\partial L}{\partial q^i}\,,
\tag{12.28}
$$

so that

$$
X_\alpha = \frac{\partial}{\partial x^\alpha} + v^i_\alpha \frac{\partial}{\partial q^i} + (X_\alpha)^i_\beta \frac{\partial}{\partial v^i_\beta}\,,
$$

that is $(X_1,\dots,X_k)$ is a SOPDE.

Theorem 12.2. *Let L be a Lagrangian and $\mathbf{X}=(X_1,\dots,X_k)$ a k-vector field such that*

$$dx^\alpha(X_\beta)=\delta^\alpha_\beta,\quad \sum_{\alpha=1}^k \iota_{X_\alpha}\Omega^\alpha_L = dE_L+\sum_{\alpha=1}^k \frac{\partial L}{\partial x^\alpha}dx^\alpha\,,$$

where $E_L=\Delta(L)-L$ and $1\leq \alpha,\beta\leq k$. Then

(1) *If L is regular, $\mathbf{X}=(X_1,\dots,X_k)$ is a SOPDE.*
Moreover, if $\psi:\mathbb{R}^k\to\mathbb{R}^k\times T^1_kQ$ is integral section of $\mathbf{X}$, then

$$\phi:\mathbb{R}^k\xrightarrow{\psi}\mathbb{R}^k\times T^1_kQ\xrightarrow{p_Q}Q$$

is a solution of the Euler-Lagrange equations (12.9).

(2) *If $(X_1,\dots,X_k)$ is integrable and $\phi^{[1]}:\mathbb{R}^k\to\mathbb{R}^k\times T^1_kQ$ is an integral section, then $\phi:\mathbb{R}^k\to Q$ is a solution of the Euler-Lagrange equations (12.9).*

Proof. (1) It is a direct consequence of the third equation in (12.27) and the third equation in (12.28).

(2) If $\phi^{[1]}$ is an integral section of $\mathbf{X}$ then from the last equation in (12.27) and the local expression (12.8) of $\phi^{[1]}$, we deduce that ϕ is a solution of the Euler-Lagrange equations (12.9). □

Therefore, equations (12.26) can be considered as a geometric version of the Euler-Lagrange field equations. From now, we will refer these equations (12.26) as ***k-cosymplectic Lagrangian equations***.

Remark 12.6. If $L:\mathbb{R}^k\times T^1_kQ\longrightarrow\mathbb{R}$ is regular, then $(dx^\alpha,\Omega^\alpha_L,V)$ is a k-cosymplectic structure on $\mathbb{R}^k\times T^1_kQ$. The Reeb vector fields $(R_L)_\alpha$ corresponding to this structure are characterized by the conditions

$$\iota_{(R_L)_\alpha}dx^\beta=\delta^\beta_\alpha\,,\quad \iota_{(R_L)_\alpha}\Omega^\beta_L=0\,,$$

and they satisfy

$$(R_L)_\alpha(E_L)=-\frac{\partial L}{\partial x^\alpha}\,.$$

Hence, if we write the k-cosymplectic Hamiltonian system (9.1) for $H=E_L$ and the k-cosymplectic manifold

$$(M=\mathbb{R}^k\times T^1_kQ,dx^\alpha,\Omega^\alpha_L,V)$$

we obtain

$$dx^\alpha(X_\beta) = \delta^\alpha_\beta, \quad \sum_{\alpha=1}^{k} \iota_{X_\alpha}\Omega^\alpha = dE_L - \sum_{\alpha=1}^{k}(R_L)_\alpha(E_L)dx^\alpha ,$$

which are equations (12.26). Therefore, the k-cosymplectic Lagrangian formalism developed in this section is a particular case of the k-cosymplectic formalism described in chapter 9. As in the Hamiltonian case, when the Lagrangian is regular one can prove that there exists a solution $(X_1, \dots, X_k)$ of the system (12.26) but this solution is not unique. ◇

Definition 12.8. A k-vector field $\mathbf{X} = (X_1, \dots, X_k) \in \mathfrak{X}^k(\mathbb{R}^k \times T^1_kQ)$ is called a ***k*-cosymplectic Lagrangian *k*-vector field** for a k-cosymplectic Hamiltonian system $(\mathbb{R}^k \times T^1_kQ, dx^\alpha, \Omega^\alpha_L, E_L)$ if $\mathbf{X}$ is a solution of (12.26). We denote by $\mathfrak{X}^k_L(\mathbb{R}^k \times T^1_kQ)$ the set of all k-cosymplectic Lagrangian k-vector fields.

Remark 12.7. If we write the equations (12.26) for the case $k = 1$, we obtain

$$dt(X) = 1 \, , \quad \iota_{X_L}\Omega_L = dE_L + \frac{\partial L}{\partial t}dt \, , \tag{12.29}$$

which are equivalent to the dynamical equations

$$dt(X) = 1 \, , \quad \iota_{X_L}\Omega_L = 0 \, ,$$

where $\Omega_L = \Omega_L + dE_L \wedge dt$ is Poincaré-Cartan 2-form Poincaré-Cartan, see [Echeverría-Enríquez, Muñoz-Lecanda and Román-Roy (1991)].

It is well known that these equations give the dynamics of the non-autonomous mechanics. ◇

12.4 The Legendre transformation and the equivalence between k-cosymplectic Hamiltonian and Lagrangian formulations of classical field theories

As in the k-symplectic case, the k-cosymplectic Hamiltonian and Lagrangian description of classical field theories are two equivalent formulations when the Lagrangian function satisfies some regularity condition. The k-cosymplectic Legendre transformation transforms one of these formalisms into the other. In this section we shall define the Legendre transformation in the k-cosymplectic approach and prove the equivalence between both Hamiltonian and Lagrangian settings. Recall that in the k-

cosymplectic approach a Lagrangian is a function defined on $\mathbb{R}^k \times T_k^1 Q$, i.e., $L\colon \mathbb{R}^k \times T_k^1 Q \to \mathbb{R}$.

Definition 12.9. Let $L : \mathbb{R}^k \times T_k^1 Q \to \mathbb{R}$ be a Lagrangian, then the ***Legendre transformation*** associated to L,

$$FL : \mathbb{R}^k \times T_k^1 Q \longrightarrow \mathbb{R}^k \times (T_k^1)^* Q$$

is defined as follows

$$FL(x, \mathrm{v}_q) = (x, [FL(x, \mathrm{v}_q)]^1, \dots, [FL(x, \mathrm{v}_q)]^k)$$

where

$$[FL(x, \mathrm{v}_q)]^\alpha(u_q) = \frac{d}{ds}\Big|_{s=0} L\left(x, v_{1q}, \dots, v_{\alpha q} + s u_q, \dots, v_{kq}\right),$$

for $1 \leq \alpha \leq k$, being $u_q \in T_q Q$ and $(x, \mathrm{v}_q) = (x, v_{1q}, \dots, v_{kq}) \in \mathbb{R}^k \times T_k^1 Q$.

Using canonical coordinates $(x^\alpha, q^i, v_\alpha^i)$ on $\mathbb{R}^k \times T_k^1 Q$ and $(x^\alpha, q^i, p_i^\alpha)$ on $\mathbb{R}^k \times (T_k^1)^* Q$, we deduce that FL is locally given by

$$\begin{aligned} FL\colon \ \mathbb{R}^k \times T_k^1 Q &\to \mathbb{R}^k \times (T_k^1)^* Q \\ (x^\alpha, q^i, v_\alpha^i) &\mapsto \left(x^\alpha, q^i, \frac{\partial L}{\partial v_\alpha^i}\right). \end{aligned} \tag{12.30}$$

The Jacobian matrix of FL is the following matrix of order $n(k+1)$,

$$\begin{pmatrix} I_k & 0 & 0 & \cdots & 0 \\ 0 & I_n & 0 & \cdots & 0 \\ \dfrac{\partial^2 L}{\partial x^\alpha \partial v_1^j} & \dfrac{\partial^2 L}{\partial q^i \partial v_1^j} & \dfrac{\partial^2 L}{\partial v_1^i \partial v_1^j} & \cdots & \dfrac{\partial^2 L}{\partial v_k^i \partial v_1^j} \\ \vdots & \vdots & \vdots & & \vdots \\ \dfrac{\partial^2 L}{\partial x^\alpha \partial v_k^j} & \dfrac{\partial^2 L}{\partial q^i \partial v_k^j} & \dfrac{\partial^2 L}{\partial v_1^i \partial v_k^j} & \cdots & \dfrac{\partial^2 L}{\partial v_k^i \partial v_k^j} \end{pmatrix}$$

where I_k and I_n are the identity matrix of order k and n respectively and $1 \leq i, j \leq n$. Thus we deduce that FL is a local diffeomorphism if and only if

$$det\left(\frac{\partial^2 L}{\partial v_\alpha^i \partial v_\beta^j}\right) \neq 0.$$

Definition 12.10. A Lagrangian function $L : \mathbb{R}^k \times T_k^1 Q \longrightarrow \mathbb{R}$ is said to be ***regular*** (resp. ***hyperregular***) if the Legendre transformation FL is a local diffeomorphism (resp. global). Otherwise, L is said to be ***singular***.

From the local expressions (12.24), (12.25) and (12.30) of Θ^α, Ω^α, Θ^α_L y Ω^α_L we deduce that the relationship between the canonical and Poincaré-Cartan forms is given by $(1 \leq \alpha \leq k)$

$$\Theta^\alpha_L = FL^*\Theta^\alpha\,, \quad \Omega^\alpha_L = FL^*\Omega^\alpha\,. \tag{12.31}$$

Consider $V = \ker((\pi_{\mathbb{R}^k})_{1,0})_*$ the vertical distribution of the bundle $(\pi_{\mathbb{R}^k})_{1,0}\colon \mathbb{R}^k \times T^1_kQ \to \mathbb{R}^k \times Q$, then one easily obtains the following characterization of a regular Lagrangian (the proof of this result can be found in [Merino (1997)]).

Prop 12.2. Let $L \in \mathcal{C}^\infty(\mathbb{R}^k \times T^1_kQ)$ be a Lagrangian function. L is regular if and only if $(dx^1, \ldots, dx^k, \Omega^1_L, \ldots, \Omega^k_L, V)$ is a k-cosymplectic structure on $\mathbb{R}^k \times T^1_kQ$.

Therefore one can state the following theorem:

Theorem 12.3. *Given a Lagrangian function $L\colon \mathbb{R}^k \times T^1_kQ \to \mathbb{R}$, the following conditions are equivalent:*

(1) *L is regular.*
(2) $\det\left(\dfrac{\partial^2 L}{\partial v^i_\alpha \partial v^j_\beta}\right) \neq 0$ *with $1 \leq i, j \leq n$ and $1 \leq \alpha, \beta \leq k$.*
(3) *FL is a local diffeomorphism.*

Now we restrict ourselves to the case of hyperregular Lagrangian. In this case the Legendre transformation FL is a global diffeomorphism and thus we can define a Hamiltonian function $H : \mathbb{R}^k \times (T^1_k)^*Q \to \mathbb{R}$ by

$$H = (FL^{-1})^*E_L = E_L \circ FL^{-1}$$

where FL^{-1} is the inverse diffeomorphism of FL.

Under these conditions, we can state the equivalence between both Hamiltonian and Lagrangian formalisms.

Theorem 12.4. *Let $L\colon \mathbb{R}^k \times T^1_kQ \to \mathbb{R}$ be a hyperregular Lagrangian then:*

(1) $\mathbf{X} = (X_1, \ldots, X_k) \in \mathfrak{X}^k_L(\mathbb{R}^k \times T^1_kQ)$ *if and only if* $(T^1_kFL)(\mathbf{X}) = (FL_*(X_1), \ldots, FL_*(X_k)) \in \mathfrak{X}^k_H(\mathbb{R}^k \times (T^1_k)^*Q)$ *where* $H = E_L \circ FL^{-1}$.
(2) *There exists a one-to-one correspondence between the set of maps $\phi\colon \mathbb{R}^k \to Q$ such that $\phi^{[1]}$ is an integral section of some $(X_1, \ldots, X_k) \in \mathfrak{X}^k_L(\mathbb{R}^k \times T^1_kQ)$ and the set of maps $\psi\colon \mathbb{R}^k \to \mathbb{R}^k \times (T^1_k)^*Q$, which are integral section of some $(Y_1, \ldots, Y_k) \in \mathfrak{X}^k_H(\mathbb{R}^k \times (T^1_k)^*Q)$, being $H = (FL^{-1})^*E_L$.*

Proof. **(1)** Given FL we can consider the canonical prolongation $T^1_k FL$ following the definition given in (5.4). Thus given a k-vector field $\mathbf{X} = (X_1, \ldots, X_k) \in \mathfrak{X}^k_L(\mathbb{R}^k \times T^1_k Q)$, one can define a k-vector field on $\mathbb{R}^k \times (T^1_k)^* Q$ by means of the following diagram

$$\begin{array}{ccc} \mathbb{R}^k \times T^1_k Q & \xrightarrow{\;FL\;} & \mathbb{R}^k \times (T^1_k)^* Q \\ {\scriptstyle \mathbf{X}}\big\downarrow & & \big\downarrow {\scriptstyle (T^1_k FL)(\mathbf{X})} \\ T^1_k(\mathbb{R}^k \times T^1_k Q) & \xrightarrow{\;T^1_k FL\;} & T^1_k(\mathbb{R}^k \times (T^1_k)^* Q) \end{array}$$

that is, for each $1 \leq \alpha \leq k$, we consider the vector field on $\mathbb{R}^k \times (T^1_k)^* Q$, $FL_*(X_\alpha)$.

We now consider the function $H = E_L \circ FL^{-1} = (FL^{-1})^* E_L$; then

$$(T^1_k FL)(\mathbf{X}) = (FL_*(X_1), \ldots, FL_*(X_k)) \in \mathfrak{X}^k_H(\mathbb{R}^k \times (T^1_k)^* Q)$$

$$dx^\alpha(FL_*(X_\beta)) = \delta^\alpha_\beta \,,$$

$$\sum_{\alpha=1}^k \iota_{FL_*(X_\alpha)} \Omega^\alpha - d\Big((FL^{-1})^* E_L\Big) + \sum_{\alpha=1}^k R_\alpha\Big((FL^{-1})^* E_L\Big) dx^\alpha = 0 \,.$$

Since FL is a diffeomorphism, the above condition is equivalent to the condition

$$dx^\alpha(X_\beta) = \delta^\alpha_\beta$$

and

$$\begin{aligned} 0 &= FL^*\Big(\sum_{\alpha=1}^k \iota_{FL_*(X_\alpha)} \Omega^\alpha - d(FL^{-1})^* E_L + \sum_{\alpha=1}^k R_\alpha\Big((FL^{-1})^* E_L\Big) dx^\alpha\Big) \\ &= \sum_{\alpha=1}^k \iota_{X_\alpha} (FL)^* \Omega^\alpha - dE_L + \sum_{\alpha=1}^k R_\alpha(E_L) dx^\alpha \\ &= \sum_{\alpha=1}^k \iota_{X_\alpha} (FL)^* \Omega^\alpha - dE_L - \sum_{\alpha=1}^k \frac{\partial L}{\partial x^\alpha} dx^\alpha \,. \end{aligned}$$

But from (12.31) this occurs if and only if $\mathbf{X} \in \mathfrak{X}^k_L(\mathbb{R}^k \times T^1_k Q)$.

Finally, observe that since FL is a diffeomorphism, $T^1_k FL$ is so also, and then all k-vector field on $\mathbb{R}^k \times (T^1_k)^* Q$ is of the type $T^1_k FL(\mathbf{X})$ for some $\mathbf{X} \in \mathfrak{X}^k(\mathbb{R}^k \times T^1_k Q)$.

(2) Let $\phi\colon \mathbb{R}^k \to Q$ be a map such that its first prolongation $\phi^{[1]}$ is an integral section of some $\mathbf{X} = (X_1, \dots, X_k) \in \mathfrak{X}^k_L(\mathbb{R}^k \times T^1_k Q)$, then the map $\psi = FL \circ \phi^{[1]}$ is an integral section of

$$T^1_k FL(\mathbf{X}) = (FL_*(X_1), \dots, FL_*(X_k))\,.$$

Since we have proved in (1) that $T^1_k FL(\mathbf{X}) \in \mathfrak{X}^k_H(\mathbb{R}^k \times (T^1_k)^* Q)$, we obtain the first part of the item (2).

The converse is proved in a similar way. Notice that any k-vector field on $\mathbb{R}^k \times (T^1_k)^* Q$ is of the form $T^1_k \mathbf{X}$ for some $\mathbf{X} \in \mathfrak{X}^k(\mathbb{R}^k \times T^1_k Q)$. Thus given $\psi\colon \mathbb{R}^k \to \mathbb{R}^k \times (T^1_k)^* Q$ integral section of any $(Y_1, \dots, Y_k) \in \mathfrak{X}^k_H(\mathbb{R}^k \times (T^1_k)^* Q)$, there exists a k-vector field $\mathbf{X} \in \mathfrak{X}^k_L(\mathbb{R}^k \times T^1_k Q)$ such that $T^1_k FL(\mathbf{X}) = (Y_1, \dots, Y_k)$. Finally, the map ψ corresponds to $\phi^{[1]}$, where $\phi = (\pi_Q)_1 \circ \psi$. □

Remark 12.8. Throughout this chapter we have developed the k-cosymplectic Lagrangian formalism on the trivial bundle $\mathbb{R}^k \times T^1_k Q\colon \mathbb{R}^k$. In [Muñoz-Lecanda, Salgado and Vilariño (2005)] we study the consequences on this theory when we consider a nonstandard flat connection on the bundle $\mathbb{R}^k \times T^1_k Q\colon \mathbb{R}^k$. This paper, [Muñoz-Lecanda, Salgado and Vilariño (2005)], is devoted to the analysis of the deformed dynamical equations and solutions, both in Hamiltonian and Lagrangian settings and we establish a characterization of the energy E_L based on variational principles. We conclude that the energy function is the only function that performs the equivalence between the Hamiltonian and Lagrangian variational principles when a nonstandard flat connection is considering. As a particular case, when $k = 1$ we obtain the results of the paper [Echeverría-Enríquez, Muñoz-Lecanda and Román-Roy (1995)]. ◇

Remark 12.9. The k-cosymplectic Lagrangian and Hamiltonian formalism of first-order classical field theories are reviewed and completed in [Rey, Román-Roy, Salgado and Vilariño (2012)], where several alternative formulations are developed. First, generalizing the construction of Tulczyjew for mechanics [Tulczyjew (1976,b)], we give a new interpretation of the classical field equations (in the multisymplectic approach this study can be seen, for instance, in [de León, Martín de Diego and Santamaría-Merino (2003)]). Second, the Lagrangian and Hamiltonian formalisms are unified by giving an extension of the Skinner-Rusk formulation on classical mechanics [Skinner and Rusk (1983)]. ◇

Chapter 13

Examples

In this chapter we shall present some physical examples which can be described using the k-cosymplectic formalism (see [Muñoz-Lecanda, Salgado and Vilariño (2009)] for more details).

13.1 Electrostatic equations

Consider the 3-cosymplectic Hamiltonian equations (9.1)

$$\begin{aligned} dx^\alpha(X_\beta) &= \delta_{\alpha\beta}, \quad 1 \leq \alpha, \beta \leq 3 \\ \sum_{\alpha=1}^{3} \iota_{X_\alpha}\Omega^\alpha &= dH - \sum_{\alpha=1}^{3} R_\alpha(H) dx^\alpha , \end{aligned} \tag{13.1}$$

where H is the Hamiltonian function given by

$$\begin{aligned} H\colon\ \mathbb{R}^3 \times (T^1_3)^*\mathbb{R} &\to \mathbb{R} \\ (x^\alpha, q, p^\alpha) &\mapsto 4\pi r(x)\sqrt{g}q + \frac{1}{2\sqrt{g}} g_{\alpha\beta} p^\alpha p^\beta , \end{aligned} \tag{13.2}$$

with $1 \leq \alpha, \beta \leq 3$ and $r(x)$ is the scalar function on $\mathbb{R}^3$ determined by (7.3), and (X_1, X_2, X_3) is a 3-vector field on $\mathbb{R}^3 \times (T^1_3)^*\mathbb{R}$.

If (X_1, X_2, X_3) is a solution of (13.1) then, from (9.2), we deduce that each X_α, with $1 \leq \alpha \leq 3$ has the local expression

$$X_\alpha = \frac{\partial}{\partial x^\alpha} + \frac{1}{\sqrt{g}} g_{\alpha\beta} p^\beta \frac{\partial}{\partial q} + (X_\alpha)^\beta \frac{\partial}{\partial p^\beta},$$

and the components $(X_\alpha)^\beta$, $1 \leq \alpha, \beta \leq 3$, satisfy the identity

$$(X_1)^1 + (X_2)^2 + (X_3)^3 = -4\pi r(x)\sqrt{g} .$$

Assume that (X_1, X_2, X_3) is an integrable 3-vector field; then, if

$$\begin{aligned} \varphi : \mathbb{R}^3 &\longrightarrow \mathbb{R}^3 \times (T^1_3)^*\mathbb{R} \\ x &\rightarrow \varphi(x) = (\psi(x), \psi^1(x), \psi^2(x), \psi^3(x)) \end{aligned}$$

is an integral section of a 3-vector field (X_1, X_2, X_3) solution of (13.1), we obtain that φ is a solution of the electrostatic equations (7.4).

13.2 The massive scalar field

Consider the Hamiltonian function $H\colon \mathbb{R}^4 \times (T^1_4)^*\mathbb{R} \to \mathbb{R}$ given by

$$H(x^1, x^2, x^3, x^4, q, p^1, p^2, p^3, p^4) = \frac{1}{2\sqrt{-g}} g_{\alpha\beta} p^\alpha p^\beta - \sqrt{-g}\left(F(q) - \frac{1}{2} m^2 q^2\right),$$

where (x^1, x^2, x^3, x^4) are the coordinates on $\mathbb{R}^4$, q denotes the scalar field ϕ and $(x^1, x^2, x^3, x^4, q, p^1, p^2, p^3, p^4)$ are the canonical coordinates on $\mathbb{R}^4 \times (T^1_4)^*\mathbb{R}$.

Consider the 4-cosymplectic Hamiltonian equation

$$\begin{aligned} dx^\alpha(X_\beta) &= \delta_{\alpha\beta}, \quad 1 \leq \alpha, \beta \leq 4 \\ \sum_{\alpha=1}^{4} \iota_{X_\alpha} \Omega^\alpha &= dH - \sum_{\alpha=1}^{4} R_\alpha(H) dx^\alpha , \end{aligned}$$

associated to the above Hamiltonian function. From (11.14) one obtains that, in natural coordinates, a 4-vector field solution of this system of equations has the following local expression (with $1 \leq \alpha \leq 4$)

$$X_\alpha = \frac{\partial}{\partial x^\alpha} + \frac{1}{\sqrt{-g}} g_{\alpha\beta} p^\beta \frac{\partial}{\partial q} + (X_\alpha)^\beta \frac{\partial}{\partial p^\beta}, \tag{13.3}$$

where the functions $(X_\alpha)^\beta \in \mathcal{C}^\infty(\mathbb{R}^4 \times (T^1_4)^*\mathbb{R})$ satisfy

$$\sqrt{-g}\Big(F'(q) - m^2 q\Big) = (X_1)^1 + (X_2)^2 + (X_3)^3 + (X_4)^4 . \tag{13.4}$$

Assume that (X_1, X_2, X_3, X_4) is an integrable 4-vector field. Let $\varphi\colon \mathbb{R}^4 \to \mathbb{R}^4 \times (T^1_4)^*\mathbb{R}$, $\varphi(x) = (x, \psi(x), \psi^1(x), \psi^2(x), \psi^3(x), \psi^4(x))$ be an integral section of a 4-vector field solution of the 4-cosymplectic Hamiltonian equation. Then from (13.3) and (13.4) one obtains

$$\begin{aligned} \frac{\partial \psi}{\partial x^\alpha} &= \frac{1}{\sqrt{-g}} g_{\alpha\beta} \psi^\beta \\ \sqrt{-g}\Big(F'(\psi) - m^2 \psi\Big) &= \frac{\partial \psi^1}{\partial x^1} + \frac{\partial \psi^2}{\partial x^2} + \frac{\partial \psi^3}{\partial x^3} + \frac{\partial \psi^4}{\partial x^4} . \end{aligned}$$

Therefore, $\psi\colon \mathbb{R}^4 \to \mathbb{R}$ is a solution of the equation

$$\sqrt{-g}\Big(F'(\psi) - m^2\psi\Big) = \sqrt{-g}\frac{\partial}{\partial x^\alpha}\left(g^{\alpha\beta}\frac{\partial\psi}{\partial t^\beta}\right),$$

that is, ψ is a solution of the scalar field equation.

Remark 13.1. Some particular cases of the scalar field equation are the following:

(1) If $F = 0$ we obtain the linear scalar field equation.
(2) If $F(q) = m^2q^2$, we obtain the Klein-Gordon equation [José and Saletan (1998)],

$$(\Box + m^2)\psi = 0 .$$

◇

For the Lagrangian counterpart, we consider again the Lagrangian (11.13).

Let $\mathbf{X} = (X_1, X_2, X_3, X_4)$ be an integrable solution of equation (12.26) for L and $k = 4$, then if $\phi\colon \mathbb{R}^4 \to \mathbb{R}$ is a solution of $\mathbf{X}$, then we obtain that ϕ is a solution of the equations:

$$\begin{aligned} 0 &= \frac{\partial^2 L}{\partial x^\alpha \partial v_\alpha}\Big|_{\phi^{[1]}(t)} - \frac{\partial^2 L}{\partial q \partial v_\alpha}\Big|_{\phi^{[1]}(t)}\frac{\partial\phi}{\partial x^\alpha} + \frac{\partial^2 L}{\partial v_\alpha \partial v_\beta}\Big|_{\phi^{[1]}(t)}\frac{\partial^2\phi}{\partial x^\alpha \partial x^\beta} - \frac{\partial L}{\partial q}\Big|_{\phi^{[1]}(t)} \\ &= \sqrt{-g}\frac{\partial}{\partial x^\alpha}\left(g^{\alpha\beta}\frac{\partial\phi}{\partial x^\beta}\right) - \sqrt{-g}(F'(\phi) - m^2\phi) \end{aligned}$$

and thus, ϕ is a solution of the scalar field equation (7.42).

13.3 Harmonic maps

Let us recall that a smooth map $\varphi\colon M \to N$ between two Riemannian manifolds (M, g) and (N, h) is called *harmonic* if it is a critical point of the energy functional E, which, when M is compact, is defined as

$$E(\varphi) = \int_M \frac{1}{2} trace_g \varphi^* h \, dv_g,$$

where dv_g denotes the measure on M induced by its metric and, in local coordinates, the expression $\frac{1}{2}\mathrm{trace}_g\varphi^* h$ reads

$$\frac{1}{2}\mathrm{trace}_g\varphi^* h = \frac{1}{2}g^{ij}h_{\alpha\beta}\frac{\partial\varphi^\alpha}{\partial x^i}\frac{\partial\varphi^\beta}{\partial x^j},$$

(g^{ij}) being the inverse of the metric matrix (g_{ij}).

This definition can be extended to the case where M is not compact by requiring that the restriction of φ to every compact domain be harmonic, (for more details see [Castrillón López,García Pérez and Ratiu (2001); Castrillón López and Marsden (2008); Eells and Lemaire (1978)]).

Now we consider the particular case $M = \mathbb{R}^k$, with coordinates (x^α). In this case, taking the Lagrangian

$$\begin{aligned} L\colon \; \mathbb{R}^k \times T^1_k N &\to \mathbb{R} \\ (x^\alpha, q^i, v^i_\alpha) &\mapsto \tfrac{1}{2} g^{\alpha\beta}(x) h_{ij}(q) v^i_\alpha v^j_\beta \end{aligned}$$

and the k-cosymplectic Euler-Lagrange equations (12.26) associated to it, we obtain the following result: if $\varphi\colon \mathbb{R}^k \to N$ is such that $\varphi^{[1]}$ is an integral section of $\mathbf{X} = (X_1, \ldots, X_k)$, being $\mathbf{X} = (X_1, \ldots, X_k)$ a solution of the geometric equation (12.26), then, φ is a solution of the Euler-Lagrange equations

$$\frac{\partial^2 \varphi^i}{\partial x^\alpha \partial x^\beta} - \Gamma^\gamma_{AB} \frac{\partial \varphi^i}{\partial x^\gamma} + \widetilde{\Gamma}^i_{jk} \frac{\partial \varphi^j}{\partial x^\alpha} \frac{\partial \varphi^k}{\partial x^\beta} = 0 \qquad 1 \le i \le n\,, \tag{13.5}$$

where $\Gamma^\gamma_{\alpha\beta}$ and $\widetilde{\Gamma}^i_{jk}$ denote the Christoffel symbols of the Levi-Civita connections of g and h, respectively.

Let us observe that these equations are the Euler-Lagrange equations associated to the energy functional E, and (13.5) can be written as

$$\mathrm{trace}_g \nabla d\varphi^* h = 0,$$

where ∇ is the connection on the vector bundle $T^*\mathbb{R}^k \otimes \varphi^*(TN)$ induced by the Levi-Civita connections on $\mathbb{R}^k$ and N (see, for example, [Eells and Lemaire (1978)]). Therefore, if $\varphi\colon \mathbb{R}^k \to N$ is a solution of (13.5), then φ is harmonic.

Remark 13.2. Some examples of harmonics maps are the following:

- Identity and constant maps are harmonic.
- In the case $k = 1$, that is, when $\varphi\colon \mathbb{R} \to N$ is a curve on N, we deduce that φ is a harmonic map if and only if it is a geodesic.
- Now, consider the case $N = \mathbb{R}$ (with the standard metric). Then $\varphi\colon \mathbb{R}^k \to \mathbb{R}$ is a harmonic map if and only if it is a harmonic function, that is, is a solution of the Laplace equation.

◇

13.4 Electromagnetic field in vacuum: Maxwell's equations

As it is well known (see [Frankel (1974)]), Maxwell's equations in $\mathbb{R}^3$, are

$$\text{(Gauss's Law)} \qquad \nabla \cdot \mathbf{E} = \rho \tag{13.6}$$

$$\text{(Ampere's Law)} \qquad \nabla \times \mathbf{B} = \mathbf{j} + \frac{\partial \mathbf{E}}{\partial t} \tag{13.7}$$

$$\text{(Faraday's Law)} \quad \nabla \times \mathbf{E} + \frac{\partial \mathbf{B}}{\partial t} = 0 \tag{13.8}$$

$$\text{(Absence of Free Magnetic Poles)} \qquad \nabla \cdot \mathbf{B} = 0\,. \tag{13.9}$$

Here, the symbols in bold represent vector quantities in $\mathbb{R}^3$, whereas symbols in italics represent scalar quantities.

The first two equations are inhomogeneous, while the other two are homogeneous. Here, ρ is the charge density, $\mathbf{E}$ is the electric field vector, $\mathbf{B}$ is the magnetic field and $\mathbf{j}$ is the current density vector, which satisfies the continuity equation

$$\frac{\partial \rho}{\partial t} + \nabla \cdot \mathbf{j} = 0\,.$$

In what follows, we consider a four-dimensional formulation of Maxwell's equations. To do that, one considers the Minkowski Space of Special Relativity. Therefore, the space-time is a 4-dimensional manifold M^4 that is just topologically $\mathbb{R}^4$. A point in space-time has coordinates (x, y, z, t) which we shall write as (x^1, x^2, x^3, x^4) instead. In this space we consider the Minkowski metric ($ds^2 = \mathrm{d}r^2 - \mathrm{d}x^2$ where $\mathrm{d}r^2$ denotes the Euclidean metric of $\mathbb{R}^3$), that is, (for simplicity we shall assume the velocity of light $c = 1$):

$$ds^2 = \mathrm{d}(x^1)^2 + \mathrm{d}(x^2)^2 + \mathrm{d}(x^3)^2 - \mathrm{d}(x^4)^2\,.$$

In the four-dimensional Minkowski space, Maxwell's equations assume an extremely compact form, which we recall now, (see [Frankel (1974); Du, Hao, Hu, Hui, Shi, Wang and Wu (2011); Warnick and P. Russer (2006)] for more details).

First, we consider the Faraday 2-form

$$\begin{aligned}\mathcal{F} = {} & E_1 \mathrm{d}x^1 \wedge \mathrm{d}x^4 + E_2 \mathrm{d}x^2 \wedge \mathrm{d}x^4 + E_3 \mathrm{d}x^3 \wedge \mathrm{d}x^4 \\ & + B_1 \mathrm{d}x^2 \wedge \mathrm{d}x^3 + B_2 \mathrm{d}x^3 \wedge \mathrm{d}x^1 + B_3 \mathrm{d}x^1 \wedge \mathrm{d}x^2\,.\end{aligned} \tag{13.10}$$

If we compute $\mathrm{d}\mathcal{F}$, we obtain that the homogeneous Maxwell equations (13.8-13.9) are equivalent to $\mathrm{d}\mathcal{F} = 0$, that is, the Faraday form is closed.

Since $\mathrm{d}\mathcal{F} = 0$ in $\mathbb{R}^4$, we must have

$$\mathcal{F} = d\mathcal{A} \tag{13.11}$$

$\mathcal{A}$ being the "potential" 1-form, which is written as

$$\mathcal{A} = A_1 \mathrm{d}x^1 + A_2 \mathrm{d}x^2 + A_3 \mathrm{d}x^3 + \Phi \mathrm{d}x^4 \in \Lambda^1(\mathbb{R}^4)\,, \tag{13.12}$$

where A_1, A_2, A_3 are the components of the magnetic vector potential and Φ is the scalar electric potential.

To develop a four-dimensional formulation of the divergence law for the electric flux density (13.6) and Ampere's law (13.7), we introduce the four-current differential form

$$\mathcal{J} = j_1 \mathrm{d}x^1 + j_2 \mathrm{d}x^2 + j_3 \mathrm{d}x^3 - \rho \mathrm{d}x^4 \in \Lambda^1(\mathbb{R}^4) \tag{13.13}$$

where j_1, j_2, j_3 are the components of the electric current and ρ is the density of electric charge.

The four-dimensional formulation of the divergence law (13.6) and Ampere's law (13.7), is

$$\delta\mathcal{M} = \mathcal{J} \tag{13.14}$$

where $\mathcal{M} \in \Lambda^2(\mathbb{R}^4)$ is the *Maxwell form* defined by $\mathcal{M} = \star\mathcal{F}$ and $\delta\colon = \star\mathrm{d}\star$ is the coderivative; here $\star\colon \Omega^k(\mathbb{R}^4) \to \Omega^{4-k}(\mathbb{R}^4)$ denotes the four-dimensional Hodge operator for Minkowski's space.

In conclusion, in a four-dimensional Minkowski's space, Maxwell's equations can be written as follows

$$\begin{aligned} \mathrm{d}\mathcal{F} &= 0\,, \\ \delta\mathcal{M} &= \mathcal{J}\,. \end{aligned} \tag{13.15}$$

Now we show that, since $\mathcal{F} = \mathrm{d}\mathcal{A}$, then the inhomogeneous equation $\delta\mathcal{M} = \mathcal{J}$ is equivalent to the Euler-Lagrange equations for some Lagrangian L.

In that case, a solution of Maxwell's equations is a 1-form $\mathcal{A}$ on the Minkowski's space, that is, $\mathcal{A}$ is a section of the canonical projection $\pi_{\mathbb{R}^4}\colon T^*\mathbb{R}^4 \cong \mathbb{R}^4 \times \mathbb{R}^4 \to \mathbb{R}^4$. Here $Q = \mathbb{R}^4$. Moreover, see [Echeverría-Enríquez and Muñoz-Lecanda (1992); Saunders (1987,b)], $\mathbb{R}^4 \times T^1_4\mathbb{R}^4$ is canonically isomorphic to $(\pi_{\mathbb{R}^4})^*T^*\mathbb{R}^4 \otimes (\pi_{\mathbb{R}^4})^*T^*\mathbb{R}^4$ via the identifications

$$\begin{aligned} \mathbb{R}^4 \times x^1_4\mathbb{R}^4 \qquad\qquad &\equiv (\pi_{\mathbb{R}^4})^*T^*\mathbb{R}^4 \otimes (\pi_{\mathbb{R}^4})^*T^*\mathbb{R}^4 \\ A^{[1]}(\mathbf{t}) = (x^j, A_i(\mathbf{t}), \frac{\partial A_i}{\partial x^j}(\mathbf{t})) \equiv &\qquad \frac{\partial A_j}{\partial x^i}(\mathbf{t})(dx^i \otimes dx^j) \end{aligned}$$

where $1 \leq i, j \leq 4$ and $A_4 = \Phi$, and $\mathcal{A}^{[1]} \colon \mathbb{R}^4 \to \mathbb{R}^4 \times T^1_4\mathbb{R}^4$ is the first prolongation of a section $\mathcal{A} \in \Lambda^1(\mathbb{R}^4)$ of $\pi_{\mathbb{R}^4}$.

Then the Lagrangian $L \colon \mathbb{R}^4 \times T^1_4\mathbb{R}^4 = (\pi_{\mathbb{R}^4})^*T^*\mathbb{R}^4 \otimes (\pi_{\mathbb{R}^4})^*T^*\mathbb{R}^4 \to \mathbb{R}$ is given by

$$L(\mathcal{A}^{[1]}) = \frac{1}{2}||\mathfrak{A}(\mathcal{A}^{[1]})|| - < \mathcal{J}, \mathcal{A} >= \frac{1}{2}||\mathrm{d}\mathcal{A}|| - < \mathcal{J}, \mathcal{A} >,$$

where $\mathfrak{A}$ is the alternating operator, and we have used the induced metric on $(\pi_{\mathbb{R}^4})^*T^*\mathbb{R}^4 \otimes (\pi_{\mathbb{R}^4})^*T^*\mathbb{R}^4$ by the metric on $\mathbb{R}^4$, see [Poor (1981)]. Here, $< \mathcal{J}, \mathcal{A} >$ denotes the scalar product in $(\mathbb{R}^4)^*$ given by the scalar product on $\mathbb{R}^4$, see [Poor (1981)],

$$< \mathcal{J}, \mathcal{A} >= j_1A_1 + j_2A_2 + j_3A_3 + \rho\Phi\,.$$

As in the above section, if we take $(x^\alpha) = (x^1, x^2, x^3, x^4)$ coordinates on $\mathbb{R}^4$, q^i are the coordinates on the fibers of $T^*\mathbb{R}^4 = \mathbb{R}^4 \times \mathbb{R}^4$ and v^i_α are the induced coordinates on the fibers of $\mathbb{R}^4 \times T^1_4\mathbb{R}^4$, then L is locally given by

$$\begin{aligned} L(x^\alpha, q^i, v^i_\alpha) = \frac{1}{2}((v^2_1 - v^1_2)^2 + (v^3_1 - v^1_3)^2 + (v^3_2 - v^2_3)^2 - (v^4_1 - v^1_4)^2 \\ - (v^4_2 - v^2_4)^2 - (v^4_3 - v^3_4)^2) \\ - j_1q^1 - j_2q^2 - j_3q^3 - \rho q^4\,. \end{aligned} \tag{13.16}$$

Remark 13.3. Let us observe that for a section $\mathcal{A} = A_1\mathrm{d}x^1 + A_2\mathrm{d}x^2 + A_3\mathrm{d}x^3 + \Phi\mathrm{d}x^4$, if $\mathcal{F} = \mathrm{d}\mathcal{A}$, we have:

$$\begin{aligned} ||\mathfrak{A}(\mathcal{A}^{[1]})|| = ||\mathrm{d}\mathcal{A}|| &= \sum_{i<j<4}(\frac{\partial A_j}{\partial x^i} - \frac{\partial A_i}{\partial x^j})^2 - \sum_{i<4}(\frac{\partial \Phi}{\partial x^i} - \frac{\partial A_i}{\partial x^4})^2 \\ &= ||B||^2 - ||E||^2\,. \end{aligned}$$

◇

Now, we consider the 4-cosymplectic equation

$$\begin{aligned} \mathrm{d}x^\alpha(X_\beta) &= \delta^\alpha_\beta, \quad 1 \leq A, B \leq 4\,, \\ \sum_{\alpha=1}^4 i_{X_\alpha}\Omega^\alpha_L &= dE_L + \sum_{\alpha=1}^4 \frac{\partial L}{\partial x^\alpha}\mathrm{d}x^\alpha \end{aligned} \tag{13.17}$$

where the Lagrangian L is given by (13.16) and $\mathbf{X} = (X_1, X_2, X_3, X_4)$ is a 4-vector field on $\mathbb{R}^4 \times T^1_4\mathbb{R}^4$.

Let $\mathcal{A} \in \Lambda^1(\mathbb{R}^4)$ be a section of $\pi_{\mathbb{R}^4}$ which is a solution of $\mathbf{X}$, then from (13.16) we obtain that $\mathcal{A}$ is a solution of the following system of equations:

$$\begin{aligned}
&\frac{\partial^2 A_1}{\partial x^2 \partial x^2} - \frac{\partial^2 A_2}{\partial x^1 \partial x^2} + \frac{\partial^2 A_1}{\partial x^3 \partial x^3} - \frac{\partial^2 A_3}{\partial x^1 \partial x^3} - \frac{\partial^2 A_1}{\partial x^4 \partial x^4} + \frac{\partial^2 \Phi}{\partial x^1 \partial x^4} = -j_1 \\
&\frac{\partial^2 A_2}{\partial x^1 \partial x^1} - \frac{\partial^2 A_1}{\partial x^2 \partial x^1} + \frac{\partial^2 A_2}{\partial x^3 \partial x^3} - \frac{\partial^2 A_3}{\partial x^2 \partial x^3} - \frac{\partial^2 A_2}{\partial x^4 \partial x^4} + \frac{\partial^2 \Phi}{\partial x^2 \partial x^4} = -j_2 \\
&\frac{\partial^2 A_3}{\partial x^1 \partial x^1} - \frac{\partial^2 A_1}{\partial x^3 \partial x^1} + \frac{\partial^2 A_3}{\partial x^2 \partial x^2} - \frac{\partial^2 A_2}{\partial x^3 \partial x^2} - \frac{\partial^2 A_3}{\partial x^4 \partial x^4} + \frac{\partial^2 \Phi}{\partial x^3 \partial x^4} = -j_3 \\
&\frac{\partial^2 A_1}{\partial x^4 \partial x^1} - \frac{\partial^2 \Phi}{\partial x^1 \partial x^1} + \frac{\partial^2 A_2}{\partial x^4 \partial x^2} - \frac{\partial^2 \Phi}{\partial x^2 \partial x^2} + \frac{\partial^2 A_3}{\partial x^4 \partial x^3} - \frac{\partial^2 \Phi}{\partial x^3 \partial x^3} = -\rho \,.
\end{aligned} \tag{13.18}$$

On the other hand, using $\mathcal{F} = \mathrm{d}\mathcal{A}$, from (13.10) one obtains that equations (13.18) can we written as follows

$$\begin{aligned}
-\frac{\partial B_3}{\partial x^2} + \frac{\partial B_2}{\partial x^3} + \frac{\partial E_1}{\partial x^4} &= -j_1 \\
\frac{\partial B_3}{\partial x^1} - \frac{\partial B_1}{\partial x^3} + \frac{\partial E_2}{\partial x^4} &= -j_2 \\
-\frac{\partial B_2}{\partial x^1} + \frac{\partial B_1}{\partial x^2} + \frac{\partial E_3}{\partial x^4} &= -j_3 \\
\frac{\partial E_1}{\partial x^1} + \frac{\partial E_2}{\partial x^2} + \frac{\partial E_3}{\partial x^3} &= \rho
\end{aligned}$$

which is equivalent to the condition $\delta\mathcal{M} = \mathcal{J}$.

In conclusion, the 4-cosymplectic equation (13.17) is a geometric version of the inhomogeneous Maxwell equation $\delta\mathcal{M} = \mathcal{J}$, and considering $\mathcal{F} := \mathrm{d}\mathcal{A}$ we also recover the homogeneous Maxwell equation $\mathrm{d}\mathcal{F} = 0$.

Remark 13.4.

(1) In the particular case $\mathcal{J} = 0$, that is when $\rho = 0, \mathbf{j} = 0$, the Lagrangian (13.16) is a function defined on $\mathcal{C}^\infty(T^1_4\mathbb{R}^4)$. Therefore, it is another example of the k-symplectic Lagrangian formalism. This Lagrangian corresponds to the electromagnetic field without currents.

(2) The Lagrangian (13.16) can also be written as follows:

$$L = -\frac{1}{4} f_{ik} f^{ik} - < \mathcal{J}, \mathcal{A} > ,$$

where

$$f_{ik} = \frac{\partial A_k}{\partial x^i} - \frac{\partial A_i}{\partial x^k} \quad \text{and} \quad f_{ik} f^{ik} = g^{il} g^{km} f_{ik} f_{lm} \,.$$

This Lagrangian can be extended, in the presence of gravitation, as follows (see [Carmeli (2001)]):

$$L = -\frac{1}{4}\sqrt{-g} f_{ik} f^{ik} - \sqrt{-g} < \mathcal{J}, \mathcal{A} > , \tag{13.19}$$

where now we have used the space-time metric tensor (g_{ij}) to raise the indices of the Maxwell tensor,

$$f^{ik} = g^{il} g^{km} f_{lm} \, .$$

In this case, in a similar way to the above discussion, and using that

$$\frac{\partial L}{\partial v^4_\beta} = \sqrt{-g} f^{4\beta} \quad , \quad \frac{\partial L}{\partial v^i_\beta} = \sqrt{-g} f^{i\beta} \, , \quad 1 \leq \beta, i \leq 3,$$

we obtain that equations (13.17) for the Lagrangian are the geometric version of the following equations

$$\begin{aligned} \nabla_k f^{4k} &= \rho \\ \nabla_k f^{ik} &= j_i, \quad i = 1,2,3 \\ \nabla_l f_{ik} + \nabla_i f_{kl} + \nabla_k f_{li} &= 0 \, , \end{aligned} \tag{13.20}$$

where

$$\nabla_k f^{ik} := \frac{1}{\sqrt{-g}} \frac{\partial}{\partial x^k} \left(\sqrt{-g} f^{ik} \right)$$

is the covariant divergence of a skew-symmetric tensor in the curved spacetime. These equations (13.20) are called the Maxwell equations in the presence of gravitation, see [Carmeli (2001)].

◇

Finally it is important to observe that all these physical examples which can be described using the k-symplectic formalism can also be described using the k-cosymplectic approach.

Chapter 14

k-symplectic Systems versus Autonomous k-cosymplectic Systems

In this book we are presenting two different approaches to describe first-order classical field theories: first, when the Lagrangian and Hamiltonian do not depend on the base coordinates, and, later, when the Lagrangian and Hamiltonian also depend on the "space-time" coordinates. However, if we observe the corresponding descriptions we see that in local coordinates they give a geometric description of the same system of partial differential equations. Therefore the natural question is: *Is there any relationship between k-symplectic and k-cosymplectic systems?* In this section we give an affirmative answer to this question. Naturally, this relation will be established only when the Lagrangian and Hamiltonian do not depend on the base coordinates.

Along this section we work over the geometrical models of k-symplectic and k-cosymplectic manifolds, that is $(T^1_k)^*Q$ and $\mathbb{R}^k \times (T^1_k)^*Q$ and the Lagrangian counterparts T^1_kQ and $\mathbb{R}^k \times T^1_kQ$. However the following results and comments can be extended to the case $\mathbb{R}^k \times M$ and M, being M an arbitrary k-symplectic manifold.

Following a similar terminology to that in mechanics, we introduce the following definition.

Definition 14.1. A k-cosymplectic Hamiltonian system $(\mathbb{R}^k \times (T^1_k)^*Q, \mathcal{H})$ is said to be ***autonomous*** if $\mathcal{L}_{R_\alpha}\mathcal{H} = {}^{\partial\mathcal{H}}/_{\partial x^\alpha} = 0$ for all $1 \leq \alpha \leq k$.

Observe that the condition in Definition 14.1 means that $\mathcal{H}$ does not depend on the variables x^α, and thus $\mathcal{H} = \bar{\pi}_2^* H$ for some $H \in \mathrm{C}^\infty((T^1_k)^*Q)$, being $\bar{\pi}_2 \colon \mathbb{R}^k \times (T^1_k)^*Q \to (T^1_k)^*Q$ the canonical projection.

For an autonomous k-cosymplectic Hamiltonian system, the equations

(9.1) become

$$\sum_{\alpha=1}^{k} \iota_{\bar{X}_\alpha} \Omega^\alpha = \mathrm{d}\mathcal{H}, \quad \eta^\alpha(\bar{X}_\beta) = \delta^\alpha_\beta \,. \tag{14.1}$$

Therefore:

Prop 14.1. Every autonomous k-cosymplectic Hamiltonian system $(\mathbb{R}^k \times (T^1_k)^*Q, \mathcal{H})$ defines a k-symplectic Hamiltonian system $((T^1_k)^*Q, H)$, where $\mathcal{H} = \bar{\pi}_2^* H$, and conversely.

We have the following result for the solutions of the HDW equations.

Theorem 14.1. *Let $(\mathbb{R}^k \times (T^1_k)^*Q, \mathcal{H})$ be an autonomous k-cosymplectic Hamiltonian system and let $((T^1_k)^*Q, H)$ be its associated k-symplectic Hamiltonian system. Then, every section $\bar{\psi}\colon \mathbb{R}^k \to \mathbb{R}^k \times (T^1_k)^*Q$, that is a solution of the HDW-equation (10.1) for the system $(\mathbb{R}^k \times (T^1_k)^*Q, \mathcal{H})$ defines a map $\psi\colon \mathbb{R}^k \to (T^1_k)^*Q$ which is a solution of the HDW-equation (4.1) for the system $((T^1_k)^*Q, H)$; and conversely.*

Proof. Since $\mathcal{H} = \pi_2^* H$ we have

$$\frac{\partial \mathcal{H}}{\partial q^i} = \frac{\partial H}{\partial q^i}, \quad \frac{\partial \mathcal{H}}{\partial p^\alpha_i} = \frac{\partial H}{\partial p^\alpha_i} \,. \tag{14.2}$$

Let $\bar{\psi}\colon \mathbb{R}^k \to \mathbb{R}^k \times (T^1_k)^*Q$ be a section of the projection $\bar{\pi}_1$, which in coordinates is expressed by $\bar{\psi}(x) = (x, \bar{\psi}^i(x), \bar{\psi}^\alpha_i(x))$. Then we construct the map $\psi = \bar{\pi}_2 \circ \bar{\psi}\colon \mathbb{R}^k \to (T^1_k)^*Q$, which in coordinates is expressed as $\psi(x) = (\psi^i(x), \psi^\alpha_i(x)) = (\bar{\psi}^i(x), \bar{\psi}^\alpha_i(x))$. Thus if $\bar{\psi}$ is a solution of the HDW-equations (10.1), from (14.2) we obtain that ψ is a solution of the HDW-equations (4.1), and the statement holds.

Conversely, consider a map $\psi\colon \mathbb{R}^k \to (T^1_k)^*Q$. We define $\bar{\psi} = (Id_{\mathbb{R}^k}, \psi)\colon \mathbb{R}^k \to \mathbb{R}^k \times (T^1_k)^*Q$. Furthermore, if $\psi(x) = (\psi^i(x), \psi^\alpha_i(x))$, then $\bar{\psi}(x) = (x, \bar{\psi}^i(x), \bar{\psi}^\alpha_i(x))$, with $\bar{\psi}^i(x) = \psi^i(x)$ and $\bar{\psi}^\alpha_i(x) = \psi^\alpha_i(x)$ (observe that, in fact, $\mathrm{Im}\,\bar{\psi} = \mathrm{graph}\,\psi$). Hence, if ψ is a solution of the HDW-equations (4.1), from (14.2) we obtain that $\bar{\psi}$ is a solution of the HDW-equations (10.1), and the statement holds. □

For k-vector fields that are solutions of the geometric field equations (3.6) and (14.1) we have:

Prop 14.2. Let $(\mathbb{R}^k \times (T^1_k)^*Q, \mathcal{H})$ be an autonomous k-cosymplectic Hamiltonian system and let $((T^1_k)^*Q, H)$ be its associated k-symplectic Hamiltonian system. Then every k-vector field $\mathbf{X} \in \mathfrak{X}^k_H(T^1_k)^*Q)$ defines a k-vector field $\bar{\mathbf{X}} \in \mathfrak{X}^k_{\mathcal{H}}(\mathbb{R}^k \times (T^1_k)^*Q)$.

Furthermore, $\mathbf{X}$ is integrable if, and only if, its associated $\bar{\mathbf{X}}$ is integrable too.

Proof. Let $\mathbf{X} = (X_1, \dots, X_k) \in \mathfrak{X}^k_H((T^1_k)^*Q)$. For every $\alpha = 1, \dots, k$, let $\bar{X}_\alpha \in \mathfrak{X}(\mathbb{R}^k \times (T^1_k)^*Q)$ be the *suspension* of the corresponding vector field $X_\alpha \in \mathfrak{X}((T^1_k)^*Q)$, which is defined as follows (see [Abraham and Marsden (1978)], p. 374, for this construction in mechanics): for every $\mathrm{p} \in (T^1_k)^*Q$, let $\gamma^\alpha_{\mathrm{p}} \colon \mathbb{R} \to (T^1_k)^*Q$ be the integral curve of X_α passing through p; then, if $x_0 = (x^1_0, \dots, x^k_0) \in \mathbb{R}^k$, we can construct the curve $\bar{\gamma}^\alpha_{\bar{\mathrm{p}}} \colon \mathbb{R} \to \mathbb{R}^k \times (T^1_k)^*Q$, passing through the point $\bar{\mathrm{p}} \equiv (x_0, \mathrm{p}) \in \mathbb{R}^k \times (T^1_k)^*Q$, given by $\bar{\gamma}^\alpha_{\bar{\mathrm{p}}}(x^\alpha) = (x^1_0, \dots, x^\alpha + x^\alpha_0, \dots, x^k_0; \gamma_{\mathrm{p}}(x^\alpha))$. Therefore, $\bar{X}_\alpha$ is the vector field tangent to $\bar{\gamma}^\alpha_{\bar{\mathrm{p}}}$ at (x_0, p). In natural coordinates, if X_α is locally given by

$$X_\alpha = (X_\alpha)^i \frac{\partial}{\partial q^i} + (X_\alpha)^\beta_i \frac{\partial}{\partial p^\beta_i}$$

then $\bar{X}_\alpha$ is locally given by

$$\bar{X}_\alpha = \frac{\partial}{\partial x^\alpha} + (\bar{X}_\alpha)^i \frac{\partial}{\partial q^i} + (\bar{X}_\alpha)^\beta_i \frac{\partial}{\partial p^\beta_i} = \frac{\partial}{\partial x^\alpha} + \bar{\pi}^*_2 (X_\alpha)^i \frac{\partial}{\partial q^i} + \bar{\pi}^*_2 (X_\alpha)^\beta_i \frac{\partial}{\partial p^\beta_i} \,.$$

Observe that the $\bar{X}_\alpha$ are $\bar{\pi}_2$-projectable vector fields, and $(\bar{\pi}_2)_* \bar{X}_\alpha = X_\alpha$. In this way we have defined a k-vector field $\bar{\mathbf{X}} = (\bar{X}_1, \dots, \bar{X}_k)$ in $\mathbb{R}^k \times (T^1_k)^*Q$. Therefore, taking (8.6) into account, we obtain

$$\sum_{\alpha=1}^k \iota_{\bar{X}_\alpha} \Omega^\alpha - \mathrm{d}\mathcal{H} = \sum_{\alpha=1}^k \iota_{\bar{X}_\alpha} \bar{\pi}^*_2 \omega^\alpha - \mathrm{d}(\bar{\pi}^*_2 H) = \bar{\pi}^*_2 \Big(\sum_{\alpha=1}^k \iota_{(\bar{\pi}_2)_* \bar{X}_\alpha} \omega^\alpha - \mathrm{d}H \Big) = 0 \,,$$

since $\mathbf{X} = (X_1, \dots, X_k) \in \mathfrak{X}^k_H (T^1_k)^*Q$ and therefore $\bar{X} = (\bar{X}_1, \dots, \bar{X}_k) \in \mathfrak{X}^k_{\mathcal{H}}(\mathbb{R}^k \times (T^1_k)^*Q)$.

Furthermore, if $\psi \colon \mathbb{R}^k \to (T^1_k)^*Q$ is an integral section of $\mathbf{X}$, then $\bar{\psi} \colon \mathbb{R}^k \to \mathbb{R}^k \times (T^1_k)^*Q$ such that $\bar{\psi} = (Id_{\mathbb{R}^k}, \psi)$ (see Theorem 14.1) is an integral section of $\bar{\mathbf{X}}$.

Now, if $\bar{\psi}$ is an integral section of $\bar{\mathbf{X}}$, the equations (10.1) hold for $\bar{\psi}(x) = (x, \bar{\psi}^i(x), \bar{\psi}^\alpha_i(x))$ and, since $(\bar{X}_\alpha)^i = \bar{\pi}^*_2 (X_\alpha)^i$ and $(\bar{X}_\alpha)^\beta_i = \bar{\pi}^*_2 (X_\alpha)^\beta_i$, this is equivalent to say that the equations (4.1) hold for $\psi(x) = (\psi^i(x), \psi^\alpha_i(x))$; in other words, ψ is an integral section of $\mathbf{X}$. □

Remark 14.1. The converse statement is not true. In fact, the k-vector fields that are solution of the geometric field equations (14.1) are not completely determined, and then there are k-vector fields in $\mathfrak{X}^k_{\mathcal{H}}(\mathbb{R}^k \times (T^1_k)^*Q)$ that are not $\bar{\pi}_2$-projectable (in fact, it suffices to take their undetermined component functions to be not $\bar{\pi}_2$-projectable). However, we have the following partial result:

Prop 14.3. Let $((T^1_k)^*Q, H)$ be an admissible k-symplectic Hamiltonian system, and we consider $(\mathbb{R}^k \times (T^1_k)^*Q,\ \mathcal{H})$ its associated autonomous k-cosymplectic Hamiltonian system. Then, every integrable k-vector field $\bar{\mathbf{X}} \in \mathfrak{X}^k_{\mathcal{H}}(\mathbb{R}^k \times (T^1_k)^*Q)$ is associated with an integrable k-vector field $\mathbf{X} \in \mathfrak{X}^k_H((T^1_k)^*Q)$.

Proof. If $\bar{\mathbf{X}} \in \mathfrak{X}^k_{\mathcal{H}}(\mathbb{R}^k \times (T^1_k)^*Q)$ is an integrable k-vector field, denote by $\bar{\mathcal{S}}$ the set of its integral sections (i.e., the solutions of the HDW-equation (10.1)). Let $\mathcal{S}$ be the set of maps $\psi \colon \mathbb{R}^k \to (T^1_k)^*Q$ associated with these sections by Theorem 14.1. But, since that $((T^1_k)^*Q, \omega^\alpha, H)$ is an admissible k-symplectic Hamiltonian system, we have that they are admissible solutions of the HDW-equation (4.1), and then they are integral sections of some integrable k-vector field $\mathbf{X} \in \mathfrak{X}^k((T^1_k)^*Q)$. Then, by Proposition 4.2, $\mathbf{X}$ satisfies the field equation (3.6) on the image of ψ, for every $\psi \in \mathcal{S}$, and thus $\mathbf{X} \in \mathfrak{X}^k_H((T^1_k)^*Q)$ since every point of $(T^1_k)^*Q$ is on the image of one of these sections. □

We now consider the Lagrangian case. In this situation we define

Definition 14.2. A k-cosymplectic (or k-precosymplectic) Lagrangian system is said to be ***autonomous*** if $\dfrac{\partial \mathcal{L}}{\partial x^\alpha} = 0$ or, what is equivalent, $\dfrac{\partial \mathcal{E}_{\mathcal{L}}}{\partial x^\alpha} = 0$.

Now, all the results obtained for the Hamiltonian case can be stated and proved in the same way for Lagrangian approach, considering the systems $(\mathbb{R}^k \times T^1_kQ, \mathcal{L})$ and (T^1_kQ, L) instead of $(\mathbb{R}^k \times (T^1_k)^*Q, \mathcal{H})$ and $((T^1_k)^*Q, H)$.

The following table summarizes the above discussion (we also include the particular case of classical mechanics).

	HAMILTONIAN FORMALISM Geometric Hamiltonian equations	LAGRANGIAN FORMALISM Geometric Lagrangian equations
k-cosymplectic formalism	$dt^A(X_B) = \delta^A_B$ $\sum_{A=1}^{k} \iota_{X_A}\omega^A = dH - \sum_{A=1}^{k} \frac{\partial H}{\partial t^A} dt^A$ $(X_1, \dots, X_k)$ k-vector field on $\mathbb{R}^k \times (T^1_k)^*Q$	$dt^A(Y_B) = \delta^A_B$ $\sum_{A=1}^{k} i_{Y_A}\omega^A_L = dE_L + \sum_{A=1}^{k} \frac{\partial L}{\partial t^A} dt^A$ $(Y_1, \dots, Y_k)$ k-vector field on $\mathbb{R}^k \times T^1_k Q$
k-symplectic formalism	$\sum_{A=1}^{k} \iota_{X_A}\omega^A = dH$ $(X_1, \dots, X_k)$ k-vector field on $(T^1_k)^*Q$	$\sum_{A=1}^{k} i_{Y_A}\omega^A_L = dE_L$ $(Y_1, \dots, Y_k)$ k-vector field on $T^1_k Q$
Cosymplectic formalism $k = 1$ (Non-autonomous mechanics)	$dt(X) = 1$ $\iota_X\omega = dH - \frac{\partial H}{\partial t} dt$ X vector field on $\mathbb{R} \times T^*Q$ or equivalently $dt(X) = 1\,,\ i_X\Omega = 0$ where $\Omega = \omega + dH \wedge dt$	$dt(Y) = 1$ $i_Y\omega_L = dE_L + \frac{\partial L}{\partial t} dt$ Y vector field on $\mathbb{R} \times TQ$ or equivalently $dt(Y) = 1\,,\ i_Y\Omega_L = 0$ where $\Omega_L = \omega_L + dE_L \wedge dt$
Symplectic formalism $k = 1$ (Autonomous mechanics)	$\iota_X\omega = dH$ X vector field on T^*Q	$i_Y\omega_L = dE_L$ Y vector field on TQ

PART 4

Relationship between k-symplectic and k-cosymplectic approaches and the multisymplectic formalism

Chapter 15

Multisymplectic Formalism

In this book, we have developed a framework to describe classical field theories using k-symplectic and k-cosymplectic manifolds. An alternative geometric framework is the multisymplectic formalism [Cariñena, Crampin and Ibort (1991); Gotay, Isenberg, Marsden and Montgomery (2004); Gotay, Isenberg, Marsden (2004); Marsden and Shkoller (1999); Román-Roy (2009)], first introduced in [Kijowski (1973); Kijowski and Szczyrba (1975); Kijowski and Tulczyjew (1979); Sniatycki (1970)], which is based on the use of multisymplectic manifolds. In particular, jet bundles are the appropriate domain to develop the Lagrangian formalism [Saunders (1989)], and different kinds of multimomentum bundles are used for developing the Hamiltonian description [Echeverría-Enríquez, Muñoz-Lecanda and Román-Roy (2000); Hélein and Kouneiher (2004); de León, Marrero and Martín de Diego (2009)]. In these models, the field equations can also be obtained in terms of multivector fields [Echeverría-Enríquez, Muñoz-Lecanda and Román-Roy (1998, 1999); Paufler and Römer (2002b)].

Multisymplectic models allow us to describe a higher variety of field theories than the k-cosymplectic or k-symplectic models, since for the latter the configuration bundle of the theory must be a trivial bundle; which is not the case for the multisymplectic formalism. The main goal of this chapter is to show the equivalence between the multisymplectic and k-cosymplectic descriptions, when theories with trivial configuration bundles are considered, for both the Hamiltonian and Lagrangian formalisms (for more details see [Rey, Román-Roy, Salgado and Vilariño (2011)]). In this way we complete the results obtained in [de León, McLean, Norris, Rey and Salgado (2002); de León, Merino, Oubiña, Rodrigues and Salgado (1998)], where an initial analysis about the relation between multisymplectic, k-cosymplectic and k-symplectic structures was carried out.

15.1 First order jet bundles

For a more detail discussion of the contents of this section, we refer to [Echeverría-Enríquez, Muñoz-Lecanda and Román-Roy (1996); Saunders (1989)].

Let $\pi : E \to M$ be a bundle where E is an $(m+n)$-dimensional manifold, which is fibered over an m-dimensional manifold M.

If (y^i) are coordinates on M, where $1 \leq i \leq m$, then we denote the fibered coordinates on E by (y^i, u^α) where $1 \leq \alpha \leq n$.

Definition 15.1. If (E, π, M) is a fiber bundle then a ***local section*** of π is a map $\phi : W \subset M \to E$, where W is an open set of M, satisfying the condition $\pi \circ \phi = id_W$. If $p \in M$ then the set of all sections of π whose domains contain p will be denoted by $\Gamma_p(\pi)$.

Definition 15.2. Define the local sections $\phi, \psi \in \Gamma_p(\pi)$ to be equivalent if $\phi(p) = \psi(p)$ and if, in some fibered coordinate system (y^i, u^α) around $\phi(p)$

$$\left.\frac{\partial \phi^\alpha}{\partial y^i}\right|_p = \left.\frac{\partial \psi^\alpha}{\partial y^i}\right|_p$$

for $1 \leq i \leq n$, $1 \leq \alpha \leq n$. The equivalence class containing ϕ is called the 1-jet of ϕ at p and is denoted $j^1_p\phi$.

Let us observe that $j^1_p\phi = j^1_p\psi$ if, and only if, $\phi_*(p) = \psi_*(p)$.

The set of all 1-jets of local sections of π has a natural structure as a differentiable manifold. The atlas which describe this structure is constructed from an atlas of fibered coordinate charts on the total space E, in much the same way that the induced atlas on the tangent bundle of TM (or on the k-tangent bundle T^1_kM) is constructed from an atlas on M.

The first jet manifold of π is the set

$$\{j^1_p\phi \,|\, p \in M\,,\ \phi \in \Gamma_p(\pi)\}$$

and is denoted $J^1\pi$. The functions π_1 and $\pi_{1,0}$ called the source and target projections respectively, are defined by

$$\begin{aligned} \pi_1 : J^1\pi &\longrightarrow M \\ j^1_p\phi &\to p \end{aligned}$$

and

$$\begin{aligned} \pi_{1,0} : J^1\pi &\longrightarrow E \\ j^1_p\phi &\to \phi(p) \end{aligned}$$

Let (U, y^i, u^α) be an adapted coordinate system on E. The induced coordinate system $(U^1, y^i, u^\alpha, u_i^\alpha)$ on $J^1\pi$ is defined on

$$U^1 = \{j_p^1\phi \, : \, \phi(p) \in U\}$$

where

$$y^i(j_p^1\phi) = y^i(p) \, , \quad u^\alpha(j_p^1\phi) = u^\alpha(\phi(p)) \, , \quad u_i^\alpha(j_p^1\phi) = \left.\frac{\partial u^\alpha \circ \phi}{\partial y^i}\right|_p \tag{15.1}$$

and are known as *derivative coordinates.*

$J^1\pi$ is a manifold of dimension $m+n(1+m)$. The canonical projections $\pi_{1,0}$ and π_1 are smooth surjective submersions.

Remark 15.1. If we consider Remarks 8.1 and 12.1 one obtains that the manifolds $\mathbb{R}^k \times (T_k^1)^*Q$ and $\mathbb{R}^k \times T_k^1Q$ are two examples of jet bundles. $\diamond$

15.2 Multisymplectic Hamiltonian formalism

15.2.1 *Multimomentum bundles*

A more complete description of the multisymplectic manifolds can be found in [Cantrijn, Ibort and de León (1996, 1999); Cariñena, Crampin and Ibort (1991); Echeverría-Enríquez, Muñoz-Lecanda and Román-Roy (2000); Giachetta, Mangiarotti and Sardanashvily (1997); Gotay (1990, 1991,b)].

Definition 15.3. The couple $(\mathcal{M}, \Omega)$, with $\Omega \in \Omega^{k+1}(\mathcal{M})$ $(2 \leq k + 1 \leq \dim \mathcal{M})$, is a ***multisymplectic manifold*** if Ω is closed and 1-nondegenerate; that is, for every $p \in \mathcal{M}$ and $X_p \in T_p\mathcal{M}$, we have that $\iota_{X_p}\Omega_p = 0$ if, and only if, $X_p = 0$.

A very important example of multisymplectic manifold is the *multicotangent bundle* $\Lambda^k Q$ of a manifold Q, which is the bundle of k-forms in Q, and is endowed with a canonical multisymplectic $(k+1)$-form. Other examples of multisymplectic manifolds which are relevant in field theory are the so-called *multimomentum bundles*: let $\pi\colon E \to M$ be a fiber bundle, ($\dim M = k$, $\dim E = n + k$), where M is an oriented manifold with volume form $\omega \in \Omega^k(M)$, and denote by (x^α, q^i) the natural coordinates in E adapted to the bundle, such that $\omega = \mathrm{d}x^1 \wedge \ldots \wedge \mathrm{d}x^k \equiv \mathrm{d}^k x$. We denote by $\Lambda_2^k E$ the bundle of k-forms on E vanishing by the action of two π-vertical vector fields. This is called the *extended multimomentum bundle*, and its canonical submersions are denoted by

$$\kappa\colon \Lambda_2^k E \to E \quad ; \quad \bar{\kappa} = \pi \circ \kappa\colon \Lambda_2^k E \to M \, .$$

We can introduce natural coordinates in $\Lambda^k_2 E$ adapted to the bundle $\pi\colon E \to M$, which are denoted by $(x^\alpha, q^i, p^\alpha_i, p)$, such that $\omega = d^k x$. Then, denoting $\mathrm{d}^{k-1}x_\alpha = \iota_{\frac{\partial}{\partial x^\alpha}} \mathrm{d}^k x$, the elements of $\Lambda^k_2 E$ can be written as $p^\alpha_i \,\mathrm{d}q^i \wedge d^{k-1}x_\alpha + p\,\mathrm{d}^k x$.

$\Lambda^k_2 E$ is a subbundle of $\Lambda^k E$, and hence $\Lambda^k_2 E$ is also endowed with canonical forms. First we have the "tautological form" $\Theta \in \Omega^k(\Lambda^k_2 E)$, which is defined as follows: let $\nu_x \in \Lambda^k_2 E$, with $x \in E$ then, for every $X_1, \ldots, X_k \in T_{\nu_x}(\Lambda^k_2 E)$, we have

$$\Theta\nu_x(X_1, \ldots, X_k) := \nu(x)(T_{\nu_x}\kappa(X_1), \ldots, T_{\nu_x}\kappa(X_k)).$$

Thus we define the multisymplectic form

$$\Omega := -\mathrm{d}\Theta \in \Omega^{k+1}(\Lambda^k_2 E)$$

and the local expressions of the above forms are

$$\Theta = p^\alpha_i \mathrm{d}q^i \wedge \mathrm{d}^{k-1}x_\alpha + p\, d^k x,\ \Omega = -\mathrm{d}p^\alpha_i \wedge \mathrm{d}q^i \wedge \mathrm{d}^{k-1}x_\alpha - \mathrm{d}p \wedge d^k x\,. \quad (15.2)$$

Consider $\pi^*\Lambda^k T^*M$, which is another bundle over E, whose sections are the π-semibasic k-forms on E, and denote by $J^1\pi^*$ the quotient $\dfrac{\Lambda^k_2 E}{\pi^*\Lambda^k T^*M}$. $J^1\pi^*$ is usually called the *restricted multimomentum bundle* associated with the bundle $\pi\colon E \to M$. Natural coordinates in $J^1\pi^*$ (adapted to the bundle $\pi\colon E \to M$) are denoted by $(x^\alpha, q^i, p^\alpha_i)$. We have the natural submersions specified in the following diagram

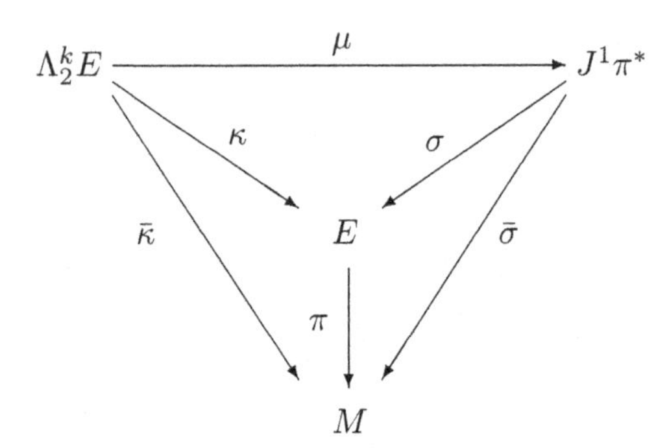

15.2.2 *Hamiltonian systems*

The Hamiltonian formalism in $J^1\pi^*$ presented here is based on the construction made in [Cariñena, Crampin and Ibort (1991)] (see also [Echeverría-Enríquez, de León, Muñoz-Lecanda and Román-Roy (2007); Echeverría-Enríquez, López, Marín-Solano, Muñoz-Lecanda and Román-Roy (2004);

Echeverría-Enríquez, Muñoz-Lecanda and Román-Roy (2000); Román-Roy (2009)]).

Definition 15.4. A section $h\colon J^1\pi^* \to \Lambda_2^k E$ of the projection μ is called a ***Hamiltonian section***. The differentiable forms $\Theta_h := h^*\Theta$ and $\Omega_h := -\mathrm{d}\Theta_h = h^*\Omega$ are called the ***Hamilton-Cartan*** k ***and*** $(k+1)$ ***forms*** of $J^1\pi^*$ associated with the Hamiltonian section h. $(J^1\pi^*, h)$ is said to be a *Hamiltonian system* in $J^1\pi^*$.

In natural coordinates we have that

$$h(x^\alpha, q^i, p_i^\alpha) = (x^\alpha, q^i, p_i^\alpha, p = -\mathcal{H}(x^\alpha, q^i, p_i^\alpha)),$$

and $\mathcal{H} \in C^\infty(U)$, $U \subset J^1\pi^*$, is a *local Hamiltonian function*. Then we have

$$\Theta_h = p_i^\alpha \mathrm{d}q^i \wedge \mathrm{d}^{k-1}x_\alpha - \mathcal{H}\mathrm{d}^k x \,,\; \Omega_h = -\mathrm{d}p_i^\alpha \wedge \mathrm{d}q^i \wedge \mathrm{d}^{k-1}x_\alpha + \mathrm{d}\mathcal{H} \wedge d^k x \,. \quad (15.3)$$

The Hamilton-De Donder-Weyl equations can also be derived from the corresponding *Hamilton–Jacobi variational principle.* In fact:

Definition 15.5. Let $(J^1\pi^*, h)$ be a Hamiltonian system. Let $\Gamma(M, J^1\pi^*)$ be the set of sections of σ. Consider the map

$$\begin{aligned} \mathbf{H}\colon\; \Gamma(M, J^1\pi^*) &\longrightarrow \mathbb{R}, \\ \psi \quad &\mapsto \textstyle\int_M \psi^*\Theta_h, \end{aligned}$$

where the convergence of the integral is assumed. The variational problem for this Hamiltonian system is the search for the critical (or stationary) sections of the functional $\mathbf{H}$, with respect to the variations of ψ given by $\psi_t = \sigma_t \circ \psi$, where $\{\sigma_t\}$ is the local one-parameter group of any compact-supported $Z \in \mathfrak{X}^{\mathrm{V}(\sigma)}(J^1\pi^*)$ (the module of σ-vertical vector fields in $J^1\pi^*$), that is:

$$\left.\frac{\mathrm{d}}{\mathrm{d}t}\right|_{t=0} \int_M \psi_t^*\Theta_h = 0.$$

The field equations for these multisymplectic Hamiltonian systems can be stated as follows

Theorem 15.1. *The following assertions on a section $\psi \in \Gamma(M, J^1\pi^*)$ are equivalent:*

1. *ψ is a critical section for the variational problem posed by the Hamilton–Jacobi principle.*
2. $\psi^* \iota_X \Omega_h = 0$, $\forall\, X \in \mathfrak{X}(J^1\pi^*)$.

If $(U; x^\alpha, q^i, p_i^\alpha)$ is a natural system of coordinates in $J^1\pi^*$, then ψ satisfies the Hamilton-De Donder-Weyl equations in U

$$\left.\frac{\partial \psi^i}{\partial x^\alpha}\right|_x = \left.\frac{\partial \mathcal{H}}{\partial p_i^\alpha}\right|_{\psi(x)}, \qquad \frac{\partial \psi_i^\alpha}{\partial x^\alpha} = -\left.\frac{\partial \mathcal{H}}{\partial q^i}\right|_{\psi(x)}, \tag{15.4}$$

where $\psi(x) = (x, \psi^i(x), \psi_i^\alpha(x))$.

15.2.3 *Relation with the k-cosymplectic Hamiltonian formalism*

In order to compare the multisymplectic and the k-cosymplectic formalisms for field theories, we consider the case when $\pi\colon E \to M$ is the trivial bundle $\mathbb{R}^k \times Q \to \mathbb{R}^k$. Then we can establish some relations between the canonical multisymplectic form on $\Lambda_2^k E \equiv \Lambda_2^k(\mathbb{R}^k \times Q)$ and the canonical k-cosymplectic structure on $\mathbb{R}^k \times (T_k^1)^*Q$.

First recall that on $M = \mathbb{R}^k$ we have the canonical volume form $\omega = \mathrm{d}x^1 \wedge \ldots \wedge \mathrm{d}x^k \equiv d^k x$. Then:

Prop 15.1. If $\pi\colon E \to M$ is the trivial bundle $\mathbb{R}^k \times Q \to \mathbb{R}^k$, we have the following diffeomorphisms:

(1) $\Lambda_2^k E \equiv \Lambda_2^k(\mathbb{R}^k \times Q)$ is diffeomorphic to $\mathbb{R}^k \times \mathbb{R} \times (T_k^1)^*Q$.
(2) $J^1\pi^* = J^{1*}(\mathbb{R}^k \times Q)$ is diffeomorphic to $\mathbb{R}^k \times (T_k^1)^*Q$.

Proof. (1) Consider the canonical embedding $\iota_x\colon Q \hookrightarrow \mathbb{R}^k \times Q$ given by $i_x(q) = (x, q)$, and the canonical submersion $\pi_Q\colon \mathbb{R}^k \times Q \to Q$. We can define the map

$$\begin{aligned} \Psi\colon\ \Lambda_2^k(\mathbb{R}^k \times Q) &\longrightarrow \mathbb{R}^k \times \mathbb{R} \times (T_k^1)^*Q \\ \nu_{(x,q)} \quad &\mapsto \quad (x, p, \nu_q^1, \ldots, \nu_q^k) \end{aligned}$$

where

$$p = \nu_{(x,q)}\left(\left.\frac{\partial}{\partial x^1}\right|_{(x,q)}, \ldots, \left.\frac{\partial}{\partial x^k}\right|_{(x,q)}\right)$$

$$\nu_q^\alpha(X) = \nu_{(x,q)}\Big(\left.\frac{\partial}{\partial x^1}\right|_{(x,q)}, \ldots, \left.\frac{\partial}{\partial x^{\alpha-1}}\right|_{(x,q)}, (\iota_x)_* X, \left.\frac{\partial}{\partial x^{\alpha+1}}\right|_{(x,q)}, \ldots, \left.\frac{\partial}{\partial x^k}\right|_{(x,q)}\Big) \qquad X \in \mathfrak{X}(Q)$$

(note that x^α and p are now global coordinates in the corresponding fibers and the global coordinate p can be identified also with the natural projection $p\colon \mathbb{R}^k \times \mathbb{R} \times (T_k^1)^*Q \to \mathbb{R}$). The inverse of Ψ is given by

$$\nu_{(x,q)} = p\,\mathrm{d}^k x|_{(x,q)} + (\pi_Q)^*_{(x,q)}\nu_q^\alpha \wedge \mathrm{d}^{k-1}x^\nu \alpha_{(x,q)}\,.$$

Thus, Ψ is a diffeomorphism. Locally Ψ is written as the identity.

(2) It is a straightforward consequence of (1), because

$$J^1\pi^* = \Lambda_2^k E/\pi^*\Lambda^k T^*M \simeq \mathbb{R}^k \times \mathbb{R} \times (T_k^1)^*Q/\mathbb{R} \simeq \mathbb{R}^k \times (T_k^1)^*Q\,.$$

□

It is important to point out that since the bundle

$$\mu\colon \Lambda_2^k E \simeq \mathbb{R}^k \times \mathbb{R} \times (T_k^1)^*Q \to J^1\pi^* \simeq \mathbb{R}^k \times (T_k^1)^*Q$$

is trivial, then the Hamiltonian sections can be taken to be global sections of the projection μ by giving a global Hamiltonian function $\mathrm{H} \in C^\infty(\mathbb{R}^k \times (T_k^1)^*Q)$.

Then we can relate the non-canonical multisymplectic form Ω_h with the k-cosymplectic structure in $\mathbb{R}^k \times (T_k^1)^*Q$ as follows:

- Starting from the forms Θ_h and Ω_h in $\mathbb{R}^k \times (T_k^1)^*Q$, we can define the forms Θ^α and Ω^α on $\mathbb{R}^k \times (T_k^1)^*Q$ as follows: for $X, Y \in \mathfrak{X}(\mathbb{R}^k \times (T_k^1)^*Q)$,

$$\Theta^\alpha(X) = -\left(\iota_{\frac{\partial}{\partial x^k}} \dots \iota_{\frac{\partial}{\partial x^1}} (\Theta_h \wedge \mathrm{d}x^\alpha)\right)(X) \tag{15.5}$$

$$\begin{aligned} \Omega^\alpha(X,Y) &= -\mathrm{d}\Theta^\alpha(X,Y) \\ &= (-1)^{k+1}\left(\iota_{\frac{\partial}{\partial x^k}} \dots \iota_{\frac{\partial}{\partial x^1}} (\Omega_h \wedge \mathrm{d}x^\alpha)\right)(X,Y) \end{aligned} \tag{15.6}$$

and the 1-forms $\eta^\alpha = \mathrm{d}x^\alpha$ are canonically defined.

- Conversely, starting from the canonical k-cosymplectic structure on $\mathbb{R}^k \times (T_k^1)^*Q$, and from $\mathcal{H}$, we can construct

$$\begin{aligned} \Theta_h &= -\mathcal{H}\mathrm{d}^k t + \Theta^\alpha \wedge \mathrm{d}^{k-1}x_\alpha\,, \\ \Omega &= -\mathrm{d}\Theta = \mathrm{d}\mathcal{H} \wedge \mathrm{d}^k t + \Omega^\alpha \wedge \mathrm{d}^{k-1}x_\alpha\,. \end{aligned} \tag{15.7}$$

Let $\mathfrak{X}_h^k(J^1\pi^*)$ be the set of k-vector fields $\mathbf{X} = (X_1, \dots, X_k)$ in $J^1\pi^*$ which are solution of the equations

$$\iota_{\mathbf{X}}\Omega_h = \iota_{X_1} \dots \iota_{X_k}\Omega_h = 0\,, \quad \iota_{\mathbf{X}}\omega = \iota_{X_1} \dots \iota_{X_k}\omega \neq 0\,, \tag{15.8}$$

(we denote by $\omega = \mathrm{d}^k x$ the volume form in M as well as its pull-backs to all the manifolds).

In a system of natural coordinates, the components of $\mathbf{X}$ are given by

$$X_\alpha = (X_\alpha)_\beta \frac{\partial}{\partial x^\alpha} + (X_\alpha)^i \frac{\partial}{\partial q^i} + (X_\alpha)^\beta \frac{\partial}{\partial v_\beta^i}\,.$$

Then, in order to assure the so-called "transversal condition" $\iota_{\mathbf{X}}\omega \neq 0$, we can take $(X_\alpha)^\beta = \delta^\beta_\alpha$, which leads to $\iota_{\mathbf{X}}\omega = 1$, and hence the other equation (15.8) give become

$$\frac{\partial \mathcal{H}}{\partial q^i} = -\sum_{\alpha=1}^{k}(X_\alpha)^\alpha_i \quad , \quad \frac{\partial \mathcal{H}}{\partial p^\alpha_i} = (X_\alpha)^i \,. \tag{15.9}$$

Let us observe that these equations coincide with equations (9.2). Thus we obtain

Theorem 15.2. *A k-vector field $\mathbf{X} = (X_1, \dots, X_k)$ on $J^1\pi^* \simeq \mathbb{R}^k \times (T^1_k)^*Q$ is a solution of equations (15.8) if, and only if, it is also a solution of the equations (9.1); that is, we have that $\mathfrak{X}^k_h(\mathbb{R}^k \times (T^1_k)^*Q) = \mathfrak{X}^k_{\mathcal{H}}(\mathbb{R}^k \times (T^1_k)^*Q)$.*

Let us observe that when $E = \mathbb{R}^k \times Q$, then, if the section $\psi : \mathbb{R}^k = M \to \mathbb{R}^k \times (T^1_k)^*Q = J^1\pi^*$ is an integral section of the k-vector field $\mathbf{X}$, ψ is a solution of the Hamilton-De Donder-Weyl equations (15.4), (as a consequence of (15.9)).

15.3 Multisymplectic Lagrangian formalism

15.3.1 *Multisymplectic Lagrangian systems*

A *Lagrangian density* is a π-semibasic k-form on $J^1\pi$, and hence it can be expressed as $\mathbb{L} = L\,\omega$, where $L \in C^\infty(J^1\pi)$ is the *Lagrangian function* associated with $\mathbb{L}$ and ω, where ω is a volume form on M. Using the canonical structures of $J^1\pi$, we can define the Poincaré-Cartan k-form $\Theta_{\mathbb{L}}$ and Poincaré-Cartan $(k+1)$-form $\Omega_{\mathbb{L}} = -d\Theta_{\mathbb{L}}$, which have the following local expressions [Echeverría-Enríquez, Muñoz-Lecanda and Román-Roy (1996)]:

$$\Theta_{\mathbb{L}} = \frac{\partial L}{\partial v^i_\alpha} \mathrm{d}q^i \wedge \mathrm{d}^{k-1}x_\alpha - \left(\frac{\partial L}{\partial v^i_\alpha} v^i_\alpha - L\right) \mathrm{d}^k x$$

$$\Omega_{\mathbb{L}} = -d\left(\frac{\partial L}{\partial v^i_\alpha}\right) \wedge \mathrm{d}q^i \wedge \mathrm{d}^{k-1}x_\alpha + d\left(\frac{\partial L}{\partial v^i_\alpha} v^i_\alpha - L\right) \wedge \mathrm{d}^k x$$

$(J^1\pi, \mathbb{L})$ is said to be a Lagrangian system.

The Lagrangian system and the Lagrangian function are *regular* if $\Omega_{\mathbb{L}}$ is a multisymplectic $(k+1)$-form. The regularity condition is locally equivalent to demand that the matrix $\left(\dfrac{\partial^2 L}{\partial v^i_\alpha \partial v^j_\beta}\right)$ be regular at every point in $J^1\pi$.

The Lagrangian field equations can be derived from a variational principle. In fact:

Definition 15.6. Let $(J^1\pi, \mathbb{L})$ be a Lagrangian system. Let $\Gamma(M,E)$ be the set of sections of π. Consider the map

$$\begin{aligned} \mathbf{L}\colon \ \Gamma(M,E) &\longrightarrow \mathbb{R}, \\ \phi \quad &\mapsto \textstyle\int_M (j^1\phi)^*\Theta_{\mathbb{L}}, \end{aligned}$$

where the convergence of the integral is assumed. The variational problem for this Lagrangian system is the search of the critical (or stationary) sections of the functional $\mathbf{L}$, with respect to the variations of ϕ given by $\phi_t = \sigma_t \circ \phi$, where $\{\sigma_t\}$ is a local one-parameter group of any compact-supported $Z \in \mathfrak{X}^{\mathrm{V}(\pi)}(E)$ (the module of π-vertical vector fields in E), that is:

$$\left.\frac{\mathrm{d}}{\mathrm{d}t}\right|_{t=0} \int_M \left(j^1\phi_t\right)^* \Theta_{\mathbb{L}} = 0.$$

This is the *Hamilton principle* of the Lagrangian formalism.

Theorem 15.3. *The following assertions on a section $\phi \in \Gamma(M,E)$ are equivalent:*

1. *ϕ is a critical section for the variational problem posed by the Hamilton principle.*
2. *$(j^1\phi)^*\iota_X\Omega_{\mathbb{L}} = 0$ for every $X \in \mathfrak{X}(J^1\pi)$, where $j^1\phi : M \to J^1\pi$ is the section defined by $j^1\phi(x) = j^1_x\phi$.*

If $\phi(x^\alpha) = (x^\alpha, \phi^i(x^\alpha))$ is a critical section then

$$j^1\phi(x^\alpha) = \left(x^\alpha, \phi^i(x^\alpha), \frac{\partial \phi^i}{\partial x^\alpha}\right)$$

satisfies the Euler-Lagrange field equations

$$\frac{\partial}{\partial x^\alpha}\left(\frac{\partial L}{\partial v^i_\alpha} \circ j^1\phi\right) - \frac{\partial L}{\partial q^i} \circ j^1\phi = 0. \tag{15.10}$$

Finally, $\Theta_{\mathbb{L}} \in \Omega^1(J^1\pi)$ being π-semibasic, we have a natural map $\widetilde{FL}\colon J^1\pi \to \mathcal{M}\pi$, given by

$$\widetilde{FL}(y) = \Theta_{\mathbb{L}}(y) \quad ; \quad y \in J^1\pi$$

which is called the ***extended Legendre transformation*** associated to the Lagrangian L. The ***restricted Legendre transformation*** is $FL =$

$\mu \circ \widetilde{FL}\colon J^1\pi \to J^1\pi^*$. Their local expressions are

$$\widetilde{FL} : (x^\alpha, q^i, v^i_\alpha) \mapsto \left(x^\alpha, q^i, \frac{\partial L}{\partial v^i_\alpha}, L - v^i_\alpha \frac{\partial L}{\partial v^i_\alpha} \right)$$

$$FL : (x^\alpha, q^i, v^i_\alpha) \mapsto \left(x^\alpha, q^i, \frac{\partial L}{\partial v^i_\alpha} \right) . \tag{15.11}$$

Moreover, we have $\widetilde{FL}^*\Theta = \Theta_{\mathbb{L}}$, and $\widetilde{FL}^*\Omega = \Omega_{\mathbb{L}}$. Observe that the Legendre transformations FL defined for the k-cosymplectic and the multisymplectic formalisms are the same, as their local expressions (12.30) and (15.11) show.

15.3.2 *Relation with the k-cosymplectic Lagrangian formalism*

Like in the Hamiltonian case, in order to compare the multisymplectic Lagrangian formalism and the k-cosymplectic Lagrangian formalism for field theories, we consider the case when $\pi\colon \mathbb{R} \to M$ is the trivial bundle $\mathbb{R}^k \times Q \to \mathbb{R}^k$. We can define the energy function E_L as

$$E_L = \Theta_{\mathbb{L}} \left(\frac{\partial}{\partial x^1}, \dots, \frac{\partial}{\partial x^k} \right)$$

whose local expression is $E_L = v^i_\alpha \dfrac{\partial L}{\partial v^i_\alpha} - L$. Then we can write

$$\Theta_{\mathbb{L}} = \frac{\partial L}{\partial v^i_\alpha} \mathrm{d}q^i \wedge d^{k-1}x_\alpha - E_L \mathrm{d}^k x \,,$$

$$\Omega_{\mathbb{L}} = -\mathrm{d} \left(\frac{\partial L}{\partial v^i_\alpha} \right) \wedge \mathrm{d}q^i \wedge d^{k-1}x_\alpha + dE_L \wedge \mathrm{d}^k x \,.$$

In this particular case, like in the Hamiltonian case, we can relate the non-canonical Lagrangian multisymplectic (or pre-multisymplectic) form $\Omega_{\mathbb{L}}$ with the non-canonical Lagrangian k-cosymplectic (or k-precosymplectic) structure in $\mathbb{R}^k \times T^1_k Q$ constructed in section 12.3.1 as follows: starting from the forms $\Theta_{\mathbb{L}}$ and $\Omega_{\mathbb{L}}$ in $J^1\pi \simeq \mathbb{R}^k \times T^1_k Q$, we can define the forms Θ^α_L and $\Omega^\alpha_L = -\mathrm{d}\Theta^\alpha_L$ on $\mathbb{R}^k \times T^1_k Q$, as follows: for $X, Y \in \mathfrak{X}(\mathbb{R}^k \times T^1_k Q)$,

$$\Theta^\alpha_L(X) = -\left(\iota_{\frac{\partial}{\partial x^k}} \dots \iota_{\frac{\partial}{\partial x^1}} (\Theta_{\mathbb{L}} \wedge \mathrm{d}x^\alpha) \right) (X)$$

$$\Omega^\alpha_L(X, Y) = (-1)^{k+1} \left(\iota_{\frac{\partial}{\partial x^k}} \dots \iota_{\frac{\partial}{\partial x^1}} (\Omega_{\mathbb{L}} \wedge \mathrm{d}x^\alpha) \right) (X, Y) \tag{15.12}$$

and the 1-forms $\eta^\alpha = \mathrm{d}x^\alpha$ are canonically defined.

Conversely, starting from the Lagrangian k-cosymplectic (or k-precosymplectic) structure on $\mathbb{R}^k \times T^1_kQ$, and from E_L, we can construct on $\mathbb{R}^k \times T^1_kQ \simeq J^1\pi$

$$\begin{aligned} \Theta_{\mathbb{L}} &= -E_L \mathrm{d}^k t + \Theta_L^\alpha \wedge d^{k-1}x_\alpha, \\ \Omega_{\mathbb{L}} &= -d\Theta_{\mathbb{L}} = dE_L \wedge d^k x + \Omega_L^\alpha \wedge d^{k-1}x_\alpha \,. \end{aligned} \tag{15.13}$$

So we have proved the following theorem, which allows us to relate the non-canonical Lagrangian multisymplectic (or pre-multisymplectic) forms with the non-canonical Lagrangian k-cosymplectic (or k-precosymplectic) structure in $\mathbb{R}^k \times T^1_kQ$.

Theorem 15.4. *The Lagrangian multisymplectic (or pre-multisymplectic) form and the Lagrangian 2-forms of the k-cosymplectic (or k-precosymplectic) structure on $J^1\pi \equiv \mathbb{R}^k \times T^1_kQ$ are related by (15.12) and (15.13).*

Let $\mathfrak{X}^k_{\mathbb{L}}(J^1\pi)$ be the set of k-vector fields $\mathbf{\Gamma} = (\Gamma_1, \ldots, \Gamma_k)$ in $J^1\pi$, that are solutions of the equations

$$\iota_{\mathbf{\Gamma}}\Omega_{\mathbb{L}} = 0 \quad , \quad \iota_{\mathbf{\Gamma}}\omega \neq 0 \,. \tag{15.14}$$

In a system of natural coordinates the components of $\mathbf{\Gamma}$ are given by

$$\Gamma_\alpha = (\Gamma_\alpha)^\beta \frac{\partial}{\partial x^\beta} + (\Gamma_\alpha)^i \frac{\partial}{\partial q^i} + (\Gamma_\alpha)^i_\beta \frac{\partial}{\partial v^i_\beta}.$$

Then, in order to assure the condition $\iota_{\mathbf{\Gamma}}\omega \neq 0$, we can take $(\Gamma_\alpha)^\beta = \delta^\beta_\alpha$, which leads to $\iota_{\mathbf{\Gamma}}\omega = 1$, and thus $\mathbf{\Gamma}$ is a solution of (15.14) if, and only if, $(\Gamma_\alpha)^i$ and $(\Gamma_\alpha)^i_\beta$ satisfy equations (12.27). When L is regular, we obtain that $(\Gamma_\alpha)^i = v^i_\alpha$, and equations (12.28) hold.

Then we can assert the following.

Theorem 15.5. *A k-vector field $\mathbf{\Gamma} = (\Gamma_1, \ldots, \Gamma_k)$ in $J^1\pi \simeq \mathbb{R}^k \times T^1_kQ$ is a solution of equations (15.14) if, and only if, it is also a solution of the equations (12.26); that is, we have that $\mathfrak{X}^k_{\mathbb{L}}(\mathbb{R}^k \times T^1_kQ) = \mathfrak{X}^k_L(\mathbb{R}^k \times T^1_kQ)$.*

Observe also that, when $E = \mathbb{R}^k \times Q$ and $J^1\pi \simeq \mathbb{R}^k \times T^1_kQ$, we have that $j^1\phi = \phi^{[1]}$, and hence, if $j^1\phi$ is an integral section of $\mathbf{\Gamma} = (\Gamma_1, \ldots, \Gamma_k)$, then ϕ is a solution to the Euler-Lagrange equations.

15.4 Correspondences between the k-symplectic, k-cosymplectic and multisymplectic formalisms

In the table on p. 185 we summarize the correspondences between the k-symplectic, k-cosymplectic and multisymplectic approaches.

HAMILTONIAN APPROACH			
	k-symplectic	k-cosymplectic	Multisymplectic
Phase space	$(T^1_k)^*Q$	$\mathbb{R}^k \times (T^1_k)^*Q$	$\Lambda^k_2 E \to J^1\pi^*$
Canonical forms	$\theta^\alpha \in \Omega^1((T^1_k)^*Q)$	$\Theta^\alpha \in \Omega^1(\mathbb{R}^k \times (T^1_k)^*Q)$	$\Theta \in \Omega^k(\Lambda^k_2 E)$
	$\omega^\alpha = -d\theta^\alpha$	$\Omega^\alpha = -d\Theta^\alpha$	$\Omega = -d\Theta$
Hamiltonian	$H : (T^1_k)^*Q \to \mathbb{R}$	$\mathcal{H} : \mathbb{R}^k \times (T^1_k)^*Q \to \mathbb{R}$	$h : J^1\pi^* \to \Lambda^k_2 E$
Geometric equations	$\sum_{\alpha=1}^{k} \iota_{X_\alpha}\omega^\alpha = dH$ $\mathbf{X} \in \mathfrak{X}^k((T^1_k)^*Q)$	$\sum_{\alpha=1}^{k} \iota_{X_\alpha}\Omega^\alpha = \mathrm{d}H - \frac{\partial H}{\partial x^\alpha}\mathrm{d}x^\alpha$ $\mathrm{d}x^\alpha(X_\beta) = \delta^\alpha_\beta$ $\mathbf{X} \in \mathfrak{X}^k(\mathbb{R}^k \times (T^1_k)^*Q)$	$\iota_{\mathbf{X}}\Omega_h = 0$ $\iota_{\mathbf{X}}\omega = 1$ $\mathbf{X} \in \mathfrak{X}^k(J^1\pi^*)$
LAGRANGIAN APPROACH			
	k-symplectic	k-cosymplectic	Multisymplectic
Phase space	T^1_kQ	$\mathbb{R}^k \times T^1_kQ$	$J^1\pi$
Lagrangian	$L : T^1_kQ \to \mathbb{R}$	$\mathcal{L} : \mathbb{R}^k \times T^1_kQ \to \mathbb{R}$	$\mathcal{L} : J^1\pi \to \mathcal{R}, \mathbb{L} = \mathcal{L}\omega$
Cartan forms	$\theta^\alpha_L \in \Omega^1(T^1_kQ)$	$\Theta^\alpha_\mathcal{L} \in \Omega^1(\mathbb{R}^k \times T^1_kQ)$	$\Theta_\mathbb{L} \in \Omega^k(J^1\pi)$
	$\omega^\alpha_L = -d\theta^\alpha_L$	$\Omega^\alpha_\mathcal{L} = -d\Theta^\alpha_\mathcal{L}$	$\Omega_\mathcal{L} = -d\Theta_\mathcal{L}$
Geometric equations	$\sum_{\alpha=1}^{k} \iota_{\Gamma_\alpha}\omega^\alpha_L = dE_L$ $\Gamma \in \mathfrak{X}^k(T^1_kQ)$	$\sum_{\alpha=1}^{k} \iota_{\Gamma_\alpha}\Omega^\alpha_\mathbb{L} = \mathrm{d}\mathcal{E}_L - \frac{\partial \mathcal{L}}{\partial x^\alpha}\mathrm{d}x^\alpha$ $\mathrm{d}x^\alpha(\Gamma_\beta) = \delta^\alpha_\beta$ $\Gamma \in \mathfrak{X}^k(\mathbb{R}^k \times T^1_kQ)$	$\iota_\Gamma\Omega_\mathbb{L} = 0$ $\iota_\Gamma\omega = 1$ $\Gamma \in \mathfrak{X}^k(J^1\pi)$

Appendix A

Symplectic Manifolds

In chapter 1 we have presented a review of the Hamiltonian mechanics on the cotangent bundle, using the canonical symplectic form on T^*Q and also the time-dependent counterpart. This approach can be extended, in a similar way, to the case of an arbitrary symplectic manifolds and cosymplectic manifolds, respectively.

Here we recall the formal definition of symplectic and cosymplectic manifolds.

The canonical model of symplectic structure is the cotangent bundle T^*Q with its canonical symplectic form.

Definition A.1. Let ω be an arbitrary 2-form on a manifold M. Then

(1) ω is called a ***presymplectic structure on*** M if ω is a closed 2-form, that is, $d\omega = 0$.
(2) ω is called an ***almost symplectic structure on*** M if it is non-degenerate.
(3) ω is called a ***symplectic structure*** if it is a closed non-degenerated 2-form.

Let us observe that if ω is an almost symplectic structure on M, then M has even dimension, say $2n$, and we have an isomorphism of $C^\infty(M)$-modules

$$\flat : \mathfrak{X}(M) \longrightarrow \bigwedge^1(M)\,, \quad \flat(Z) = \iota_Z\omega\,.$$

Let $(x^1, \ldots, x^n, y^1, \ldots, y^n)$ be the standard coordinates on $\mathbb{R}^{2n}$. The canonical symplectic form on $\mathbb{R}^{2n}$ is

$$\omega_0 = dx^1 \wedge dy^1 + \ldots + dx^n \wedge dy^n\,.$$

The most important theorem in Symplectic Geometry is the following.

Theorem A.1 (Darboux Theorem). *Let ω be an almost symplectic 2-form on a $2n$-dimensional manifold M. Then $d\omega = 0$ (that is, ω is symplectic), if and only if for each $x \in M$ there exists a coordinate neighborhood U with local coordinates $(x^1, \ldots, x^n, y^1, \ldots, y^n)$ such that*

$$\omega = \sum_{i=1}^{n} dx^i \wedge dy^i \,.$$

Taking into account this result, one could develop the Hamiltonian formulation described in section 1.1.3 substituting the cotangent bundle T^*Q by an arbitrary symplectic manifold, see [Abraham and Marsden (1978); Arnold (1978); Godbillon (1969); Godstein, Poole Jr. and Safko (2001); Holm, Schmah and Stoica (2009); Holm (2008)].

Appendix B

Cosymplectic Manifolds

Definition B.1. A ***cosymplectic manifold*** is a triple (M, η, ω) consisting of a smooth $(2n+1)$-dimensional manifold M equipped with a closed 1-form η and a closed 2-form ω, such that $\eta \wedge \omega^n \neq 0$.

In particular, $\eta \wedge \omega^n$ is a volume form on M.

The standard example of a cosymplectic manifold is provided by the extended cotangent bundle $(\mathbb{R} \times T^*N, dt, \pi^*\omega_N)$ where $t\colon \mathbb{R} \times T^*N \to N$ and $\pi\colon \mathbb{R} \times T^*N \to T^*N$ are the canonical projections and ω_N is the canonical symplectic form on T^*N.

Consider the vector bundle homomorphism

$$\begin{aligned} \flat\colon TM &\to T^*M \\ v &\mapsto \flat(v) = \iota_v\omega + (\iota_v\eta)\eta\,. \end{aligned}$$

Then $\flat$ is a vector bundle isomorphism with inverse $\sharp$. Of course, the linear homomorphism

$$\flat_x\colon T_xM \to T_x^*M$$

induced by $\flat$ is also an isomorphism, for all $x \in M$.

Given a cosymplectic manifold (M, η, ω), then there exists a distinguished vector field $\mathcal{R}$ (called the ***Reeb vector field***) such that

$$\iota_{\mathcal{R}}\eta = 1, \quad \iota_{\mathcal{R}}\omega = 0\,,$$

or, in other form,

$$\mathcal{R} = \sharp(\eta)\,.$$

Theorem B.1 (Darboux theorem for cosymplectic manifolds). *Given a cosymplectic manifold (M, η, ω), there exists, around each point x of M, a coordinate neighborhood with coordinates (t, q^i, p_i), $1 \leq i \leq n$, such that*

$$\eta = dt, \quad \omega = dq^i \wedge dp_i \,.$$

These coordinates are called ***Darboux coordinates****.*

In Darboux coordinates, we have $\mathcal{R} = \partial/\partial t$.

Using the isomorphisms $\flat$ and $\sharp$ one can associate with every function $f \in \mathcal{C}^\infty(M)$ these following vector fields:

- The ***gradient vector field***, $\operatorname{grad} f$ defined by

$$\operatorname{grad} f = \sharp(df)\,,$$

  or, equivalently,

$$\iota_{\operatorname{grad} f}\eta = \mathcal{R}(f), \quad \iota_{\operatorname{grad} f}\omega = df - \mathcal{R}(f)\eta \,.$$

- The ***Hamiltonian vector field*** X_f defined by

$$X_f = \sharp(df - \mathcal{R}(f)\eta)\,,$$

  or, equivalently,

$$\iota_{X_f}\eta = 0, \quad \iota_{X_f}\omega = df - \mathcal{R}(f)\eta \,.$$

- The ***evolution vector field*** E_f defined by

$$E_f = \mathcal{R} + X_f\,,$$

  or, equivalently,

$$\iota_{E_f}\eta = 1, \quad \iota_{E_f}\omega = df - \mathcal{R}(f)\eta \,.$$

In Darboux coordinates we have

$$\operatorname{grad} f = \frac{\partial f}{\partial t}\frac{\partial}{\partial t} + \frac{\partial f}{\partial p_i}\frac{\partial}{\partial q^i} - \frac{\partial f}{\partial q^i}\frac{\partial}{\partial p_i}\,,$$

$$X_f = \frac{\partial f}{\partial p_i}\frac{\partial}{\partial q^i} - \frac{\partial f}{\partial q^i}\frac{\partial}{\partial p_i}\,,$$

$$E_f = \frac{\partial}{\partial t} + \frac{\partial f}{\partial p_i}\frac{\partial}{\partial q^i} - \frac{\partial f}{\partial q^i}\frac{\partial}{\partial p_i}\,.$$

Consider now an integral curve $c(s) = (t(s), q^i(s), p_i(s))$ of the evolution vector field E_f: this implies that $c(s)$ should satisfy the following system of differential equations

$$\frac{dt}{ds} = 1, \quad \frac{dq^i}{ds} = \frac{\partial f}{\partial p_i}, \quad \frac{dp_i}{ds} = -\frac{\partial f}{\partial q^i}.$$

Since $\frac{dt}{ds} = 1$ implies $t(s) = s + constant$, we deduce that

$$\frac{dq^i}{dt} = \frac{\partial f}{\partial p_i}, \quad \frac{dp_i}{dt} = -\frac{\partial f}{\partial q^i},$$

since t is an affine transformation of s.

As in symplectic Hamiltonian mechanics, we can define a Poisson bracket. Indeed, if $f, g \in C^\infty(M)$, then

$$\{f, q\} = \omega(\operatorname{grad} f, \operatorname{grad} g)$$

such that we obtain the usual expression for this Poisson bracket

$$\{f, g\} = \frac{\partial f}{\partial p_i}\frac{\partial g}{\partial q^i} - \frac{\partial f}{\partial q^i}\frac{\partial g}{\partial p_i}.$$

Observe that a cosymplectic manifold is again a Poisson manifold when it is equipped with this bracket $\{\cdot, \cdot\}$.

For more details of cosymplectic manifolds see, for instance [Cantrijn, de León and Lacomba (1992); Chinea, de León and Marrero (1991)].

Appendix C

Glossary of Symbols

$V, W, \ldots$	Vector spaces
$Q, M, N, \ldots$	$\mathcal{C}^\infty$ finite-dimensional manifolds
$\mathfrak{X}(M)$	Set of vector fields on M
$\mathfrak{X}^k(M)$	Set of k-vector fields on M
$\Lambda^l M \to M$	Bundle of l-forms
$\tau\colon TQ \to Q$	Tangent bundle
$\pi\colon T^*Q \to Q$	Cotangent bundle
(q^i)	local coordinate system on Q
(q^i, p_i)	local coordinate system on T^*Q
(q^i, v^i)	local coordinate system on TQ
$f\colon U \subset N \to M$	smooth ($\mathcal{C}^\infty$) mapping
f_* or Tf	Tangent map to $f\colon M \to N$
d	Exterior derivative
ι_v	Inner product
$\mathcal{L}_X$	Lie derivative
θ	Liouville 1-form
ω	Canonical symplectic form
H	Hamiltonian function
X_H	Hamiltonian vector field
$(X_q)^v_{v_q}$	Vertical lift of X_q to TQ at v_q
Δ	Liouville vector field
J	Vertical endomorphism
Γ	Second order differential equation
L	Lagrangian function
E_L	Energy
FL	Legendre transformation

θ_L	Pullback of θ by FL
ω_L	Pullback of ω by FL
$\pi^k \colon (T^1_k)^*Q \to Q$	Cotangent bundle of k^1-covelocities
$\pi^{k,\alpha} \colon (T^1_k)^*Q \to T^*Q$	Projection over α copy of T^*Q
$\tau^k \colon T^1_kQ \to Q$	Tangent bundle of k^1-velocities
(q^i, p^α_i)	local coordinate system on $(T^1_k)^*Q$
(q^i, v^i_α)	local coordinate system on T^1_kQ
$\phi^{(1)}$	First prolongation of maps to T^1_kQ
$\phi^{[1]}$	First prolongation to $\mathbb{R}^k \times T^1_kQ$
$\{J^1, \ldots, J^k\}$	k-tangent structure
Z^C	Complete lift of a vector field Z

Bibliography

Abraham, R. and Marsden, J. E. (1978). ***Foundations of mechanics.*** 2nd edn. (Benjamin/Cummings Publishing Co., Inc., Advanced Book Program, Reading).

Albert, C. (1989). Le théoreme de réduction de Marsden-Weinstein en géometrie cosymplectique et de contact. ***J. Geom. Phys.*** **6**, 4, pp. 627-649.

Arnold, V. I. (1978). ***Mathematical Methods of classical mechanics.*** Graduate Texts in Mathematics, 60. (Springer-Verlag, New York-Heidelberg).

Arnold, L. (1998). ***Random Dynamical Systems.*** (Springer Monographs in Mathematics Springer-Verlag, Berlin).

Awane, A. (1992). k-symplectic structures. ***J. Math. Phys.*** **33**, 12, pp. 4046-4052.

Awane, A. (1994). G-espaces k-symplectiques homogènes. ***J. Geom. Phys.*** **13**, 2, pp. 139-157.

Awane, A. and Goze, M. (2000). ***Pfaffian Systems, k-symplectic Systems.*** (Kluwer Academic Publishers, Dordrecht).

Barbero-Liñan, M., de León, M. and Martín de Diego, D. (2013). Lagrangian submanifolds and Hamilton-Jacobi equation. ***Monatsh. Math.*** **171**, 3-4, pp. 269-290.

Barone, A., Esposito, F., Magee, C. J. and Scott, A. C. (1971). Theory and applications of the Sine-Gordon equation. ***Riv. Nuovo Cim.*** **171**, 2, pp. 227–267.

Bates, L. and Sniatycki, J. (1993). Nonholonomic reduction. ***Rep. Math. Phys.*** **32**, 1, pp. 99-115.

Bertin, M. C., Pimentel, B. M. and Pompeia, P. J. (2008). Hamilton-Jacobi approach for first order actions and theories with higher order derivatives. ***Annals Phys.*** **323**, pp. 527–547.

Binz, E., de León, M., Martín de Diego, D. and Socolescu, D. (2002). Nonholonomic Constraints in classical field theories. XXXIII Symposium on Mathematical Physics (Torún, 2001). ***Rep. Math. Phys.*** **49** 2-3, pp 151-166.

Bishop, A. R. and Schneider, T. (eds.) (1978). Solitons and Condensed Matter Physics. ***Proc. of the Symposium on Nonlinear (Soliton) Structure and Dynamics in Condensed Matter (Oxford, June 27-29, 1978).***

(Springer-Verlag, Berlin-New York).

Bloch, A. M. (2003). ***Nonholonomic mechanics and Control.*** Interdisciplinary Applied Mathematics Series, 24. Systems and Control. (Springer-Verlag, New York).

Bloch, A. M., Krishnaprasad, P. S., Marsden, J. E. and Murray, R. M. (1996). Nonholonomic mechanical systems with symmetry. ***Arch. Rational Mech. Anal.*** **136** 1, pp. 21-99.

Bruno, D. (2007). Constructing a class of solutions for the Hamilton-Jacobi equations in field theory. ***J. Math. Phys.*** **48**, 11, pp. 112902.

Cannas da Silva, A. and Weinstein, A. (1999). **Geometric Models for Noncommutative Algebras**. Berkeley Mathematics Lecture Notes 10. American Mathematical Society, Providence, RI; Berkeley Center for Pure and Applied Mathematics, Berkeley, CA.

Cantrijn, F., Ibort, A. and de León, M. (1996). Hamiltonian structures on multisymplectic manifolds. Geometrical structures for physical theories, I (Vietri, 1996). ***Rend. Sem. Mat. Univ. Politec. Torino*** **54**, 3, pp. 225-236.

Cantrijn, F., Ibort, A. and de León, M. (1999). On the geometry of multisymplectic manifolds. ***J. Austral. Math. Soc. Ser. A*** **66** 3, pp. 303-330.

Cantrijn, F., de León, M. and Lacomba, E. (1992). Gradient vector fields on cosymplectic manifolds. ***J. Phys. A*** **25** 1, pp. 175-188.

Carmeli, M. (2001). ***Classical Fields: General Relativity and Gauge Theory.*** (World Scientific Publishing Co., Inc., River Edge, New Jersey).

Cariñena, J. F., Crampin, M. and Ibort, L. A. (1991). On the multisymplectic formalism for first order field theories. ***Differential Geom. Appl.*** **1**, 4, pp. 345-374.

Cariñena, J. F., Gràcia, X., Marmo, G., Martínez, E., Muñoz-Lecanda, M. C. and Román-Roy, N. (2006). Geometric Hamilton-Jacobi theory. ***Int. J. Geom. Methods Mod. Phys.*** **3** 7, pp. 1417-1458.

Cariñena, J. F., Gràcia, X., Marmo, G., Martínez, E., Muñoz-Lecanda, M. C. and Román-Roy, N. (2010). Geometric Hamilton-Jacobi theory for nonholonomic dynamical systems. ***Int. J. Geom. Methods Mod. Phys.*** **7**, 3, pp. 431-454.

Castrillón López, M., García Pérez, P. L. and Ratiu, T. S. (2001). Euler-Poincaré reduction on principal bundles. ***Lett. Math. Phys.*** **58**, 2, pp. 167-180.

Castrillón López, M. and Marsden, J. E. (2008). Covariant and dynamical reduction for principal bundle field theories. ***Ann. Global Anal. Geom.*** **34**, 3, pp. 263-285.

Chinea, D., de León, M. and Marrero, J. C. (1991). Locally conformal cosymplectic manifolds and time-dependent Hamiltonian systems. ***Comment. Math. Univ. Carolin.*** **32**, 2, pp. 383-387.

Cortés, J. (2002). **Geometric, Control and Numerical Aspects of Nonholonomic Systems**. Lecture Notes in Mathematics, 1793. (Springer-Verlag, Berlin).

Cortés, J., de León, M., Martín de Diego, D. and Martínez, S. (2003). Geometric description of vakonomic and nonholonomic dynamics. Comparison of soluciones. ***SIAM J. Control Optim.*** **41**, 5, pp. 1389–1412.

Cortés, J., de León, M., Marrero, J. C., Martín de Diego, D. and Martinez, E. (2006). A survey of Lagrangian mechanics and control on Lie algebroids and groupoids. ***Int. J. Geom. Methods Mod. Phys.*** **3**, 3, pp. 509-558.

Cortés, J. Martínez, S. and Cantrijn, F. (2002). Skinner-Rusk approach to time-dependent mechanics. ***Phys. Lett. A*** **300**, 2-3, pp. 250-258.

Crampin, M. (1983). Tangent bundle geometry for Lagrangian dynamics. ***J. Phys. A*** **16**, pp. 3755-3772.

Crampin, M. (1983). Defining Euler-Lagrange fields in terms of almost tangent structures. ***Phys. Lett. A*** **95**, 9, pp. 466-468.

Davydov, A. S. (1985). ***Solitons in Molecular Systems.*** Mathematics and its Applications (Soviet Series), 4. (D. Reidel Publishing Co., Dordrecht).

Dieudonné, J. (1969). ***Foundations of modern analysis.*** 2nd edn. Pure and Applied Mathematics, Vol. 10-I. (Academic Press, New York-London).

Du, S., Hao, C., Hu, Y., Hui, Y., Shi, Q., Wang, L. and Wu, Y. (2011). Maxwell electromagnetic theory from a viewpoint of differential forms. `http://arxiv.org/abs/0809.0102v4`.

Durand, E. (1964). ***Électrostatique, les distributions.*** (Mason et Cie Editeurs, Paris).

Echeverría-Enríquez, A., De León, M., Muñoz-Lecanda, M. C. and Román-Roy, N. (2007). Extended Hamiltonian Systems in Multisymplectic field theories. ***J. Math. Phys.*** **48**, 11, p. 112901.

Echeverría-Enríquez, A., López, C. Marín-Solano, J., Muñoz-Lecanda, M. C. and Román-Roy, N. (2004). Lagrangian-Hamiltonian unified formalism for field theory. ***J. Math. Phys.*** **45**, 1, pp. 360-380.

Echeverría-Enríquez, A. and Muñoz-Lecanda, M. C. (1992). Variational calculus in several variables: a Hamiltonian approach. ***Ann. Inst. H. Poincaré Phys. Théor.*** **56**, 1, pp. 27-47.

Echeverría-Enríquez, A., Muñoz-Lecanda, M. C. and Román-Roy, N. (1991). Geometrical setting of time-dependent regular systems: Alternative models. ***Rev. Math. Phys.*** **3**, 3, pp. 301-330.

Echeverría-Enríquez, A., Muñoz-Lecanda, M. C. and Román-Roy, N. (1995). Non-standard connections in classical mechanics. ***J. Phys. A*** **28**, 19, pp. 5553-5567.

Echeverría-Enríquez, A., Muñoz-Lecanda, M. C. and Román-Roy, N. (1996). Geometry of Lagrangian first-order classical field theories. ***Fortschr. Phys.*** **44**, 3, pp. 235-280.

Echeverría-Enríquez, A., Muñoz-Lecanda, M. C. and Román-Roy, N. (1998). Multivector fields and connections: setting Lagrangian equations in field theories. ***J. Math. Phys.*** **39**, 9, pp. 4578-4603.

Echeverría-Enríquez, A., Muñoz-Lecanda, M. C. and Román-Roy, N. (1999). Multivector field formulation of Hamiltonian field theories: equations and symmetries. ***J. Phys. A*** **32**, 48, pp. 8461-8484.

Echeverría-Enríquez, A., Muñoz-Lecanda, M. C. and Román-Roy, N. (2000).

Geometry of multisymplectic Hamiltonian first-order field theories. ***J. Math. Phys.* 41**, 11, pp. 7402-7444.

Ehresmann, Ch. (1951). Les prolongements d'une variété différentiable. I. Calcul des jets, prolongement principal. ***C. R. Acad. Sci. Paris* 233**, pp. 598-600.

Eells, J. and Lemaire, L. (1978). A report on harmonic maps. ***Bull. London Math. Soc.* 10**, 1, pp. 1-68.

Frankel, T. (1974). Maxwell's equations. ***Amer. Math. Monthly* 81**, pp. 343-349.

García, P. L. and Pérez-Rendón, A. (1969). Symplectic approach to the theory of quantized fields. I. ***Comm. Math. Phys.* 13**, pp. 24-44.

García, P. L. and Pérez-Rendón, A. (1971). Symplectic approach to the theory of quantized fields. II. ***Arch. Rational Mech. Anal.* 43**, pp. 101-124.

Giachetta, G., Mangiarotti, L. and Sardanashvily, G. (1997). ***New Lagrangian and Hamiltonian Methods in field theory.*** (World Scientific Publishing Co., Inc., River Edge, NJ).

Giachetta, G., Mangiarotti, L. and Sardanashvily, G. (1999). Covariant Hamilton equations for field theory. ***J. Phys. A* 32**, 38, pp. 6629-6642.

Gibbon, J. D., James, I. N. and Moroz, I. M. (1979). The Sine-Gordon Equation as a Model for a Rapidly Rotating Baroclinic Fluid. ***Phys. Scripta* 20**, 3-4, pp. 402-408.

Ginzburg, V. L. and Landau, L. D. (1950). On the theory of superconductivity, ***Zh. Eksp. Teor. Fiz.* 20**, pp. 1064–1082.

Godbillon, C. (1969). ***Géométrie Différentielle et Mécanique Analytique.*** (French) (Hermann, Paris).

Godstein, H., Poole Jr., C. P. and Safko, J. L. (2001). ***Classical Mechanics***, 3rd edn. (Addison-Wesley Pub. Co.).

Goldschmidt, H. and Sternberg, S. (1973). The Hamilton-Cartan formalism in the calculus of variations. ***Ann. Inst. Fourier (Grenoble)* 23**, 1, pp. 203-267.

Gotay, M. J. (1990). An exterior differential systems approach to the Cartan form in ***Symplectic Geometry and Mathematical Physics (Aix-en-Provence, 1990)*** (Progr. Math., 99 (Birkhäuser Boston, Boston), pp. 160-188.

Gotay, M. J. (1991). A multisymplectic framework for classical field theory and the calculus of variations. I. Covariant Hamiltonian formalism. ***Mechanics, Analysis and Geometry: 200 Years after Lagrange.*** (North-Holland Delta Ser., North-Holland, Amsterdam), pp. 203-235.

Gotay, M. J. (1991). A multisymplectic framework for classical field theory and the calculus of variations. II. Space + time decomposition. ***Differential Geom. Appl.* 1**, 4, pp. 375-390.

Gotay, M. J., Isenberg, J., Marsden, J. E. and Montgomery, R. (2004). ***Momentum Maps and classical Relativistic fields. Part I: Covariant field theory.*** `http://arxiv.org/abs/physics/9801019v2`.

Gotay, M. J., Isenberg, J. and Marsden, J. E. (2004). Momentum Maps and classical Relativistic fields. Part II: Canonical Analysis of field theories. `http://arxiv.org/abs/math-ph/0411032`.

Grifone, J. (1972). Structure presque-tangente et connexions. I. ***Ann. Inst. Fourier (Grenoble)*** **22**, 1, pp. 287-334.

Grifone, J. (1972). Structure presque-tangente et connexions. II. ***Ann. Inst. Fourier (Grenoble)*** **22**, 3, pp. 291-338.

Grifone, J. and Mehdi, M. (1999). On the geometry of Lagrangian mechanics with non-holonomic constraints. ***J. Geom. Phys.*** **30**, 3, pp. 187-203.

Günther, C. (1987). The polysymplectic Hamiltonian formalism in field theory and calculus of variations. I. The local case. ***J. Differential Geom.*** **25**, 1, pp. 23-53.

Guo, Z. and Schmidt, I. (2012). Converting classical theories to Quantum theories by Solutions of the Hamilton-Jacobi Equation. ***Phys. Rev. D*** **86**, 045012.

Hélein, F. and Kouneiher, J. (2004). Covariant Hamiltonian formalism for the calculus of variations with several variables: Lepage-Dedecker versus De Donder-Weyl. ***Adv. Theor. Math. Phys.*** **8**, 3, pp. 565-601.

Higgins P. J. and Mackenzie, K. (1990). Algebraic constructions in the category of Lie algebroids. ***J. Algebra*** **129**, 1, pp. 194-230.

Holm, D. D., Schmah, T. and Stoica, C. (2009). Geometric mechanics and symmetry. From finite to infinite dimensions. With solutions to selected exercises by David C. P. Ellis. Oxford Texts in Applied and Engineering Mathematics, 12. (Oxford University Press, Oxford).

Holm, D. D. (2008). ***Geometric mechanics. Part I. Dynamics and symmetry.*** (Imperial College Press, London).

Infeld, E. and Rowlands, G. (2000). ***Nonlinear Waves, Solitons and Chaos.*** 2nd edn (Cambridge University Press, Cambridge).

José, J. V. and Saletan, E. J. (1998). ***Classical Dynamics. A Contemporary Approach.*** (Cambridge University Press, Cambridge).

Kanatchikov, I. V. (1998). Canonical structure of classical field theory in the polymomentum phase space. ***Rep. Math. Phys.*** **41**, 1, pp. 49-90.

Kijowski, J. (1973). A finite-dimensional canonical formalism in the classical field theory. ***Comm. Math. Phys.*** **30**, pp. 99-128.

Kijowski, J. and Szczyrba, W. (1975). Multisymplectic manifolds and the geometrical construction of the Poisson brackets in the classical field theory. Géométrie symplectique et physique mathématique (Colloq. Internat. C.N.R.S., Aix-en-Provence, 1974). Éditions Centre Nat. Recherche Sci., Paris, pp. 347–349.

Kijowski, J. and Tulczyjew, W. M. (1979). ***A Symplectic Framework for Field Theories.*** Lecture Notes in Physics, 107. (Springer-Verlag, Berlin-New York).

Klein, J. (1962). Espaces variationelles et mécanique. ***Ann. Inst. Fourier*** **12**, pp. 1–124.

Kolář, I., Michor, P. and Slovák, J. (1993). ***Natural Operations in Differential Geometry.*** (Springer-Verlag, Berlin).

Kobayashi, S. and Nomizu, K. (1963). ***Foundations of Differential Geometry. Vol I.*** (Interscience Publishers, a division of John Wiley & Sons, New York-Lond).

Lee, J. M. (2003) ***Introduction to Smooth Manifolds***. Graduate Texts in Mathematics, 218. (Springer-Verlag, New York).

de León, M., Iglesias-Ponte, D. and Martín de Diego, D. (2008). Towards a Hamilton-Jacobi theory for nonholonomic mechanical systems. ***J. Phys. A*41**, 1, pp. 015205.

de León, M., Marín, J. and Marrero, J. C. (1995). Ehresmann Connections in classical field theories in Differential geometry and its applications (Granada, 1994). ***An. Fís. Monogr.*** **2** CIEMAT, Madrid, pp. 73–89.

de León, M., Marín, J. and Marrero, J. C. (1996). A geometrical approach to classical field theories: a constraint algorithm for singular theories. ***New Developments in Differential Geometry Mathematics and Its Applications*** **350**, pp. 291–312.

de León, M., Marrero, J. C. and Martín de Diego, D. (1997). Mechanical systems with nonlinear constraints. ***Int. J. of Theor. Phys.*** **36**, 4, pp. 979–995.

de León, M., Marrero, J. C. and Martín de Diego, D. (1997). Non-holonomic Lagrangian systems in jet manifolds. ***J. Phys. A*** **30**, 4, pp. 1167-1190.

de León, M., Marrero, J. C. and Martín de Diego, D. (2003). A new geometrical setting for classical field theories. ***Banach Center Publications Institute of Mathematics, Polish Acad. Science*** **59**, pp. 189–209.

de León, M., Marrero, J. C. and Martín de Diego, D. (2009). A geometric Hamilton-Jacobi theory for classical field theories in ***Variations, Geometry and Physics***. (Nova Sci. Publ., New York,) pp. 129-140.

de León, M., Marrero, J. C. and Martín de Diego, D. (2010). Linear almost Poisson structures and Hamilton-Jacobi equation. Applications to nonholonomic mechanics. ***J. Geom. Mech.*** **2**, 2, pp. 159-198.

de León, M., Marrero, J. C., Martín de Diego, D., Salgado, M. and Vilariño, S. (2010). Hamilton-Jacobi theory in k-symplectic field theories. ***Int. J. Geom. Methods Mod. Phys.*** **7**, 8, pp. 1491-1507.

de León, M., Marrero, J. C. and Martínez, E. (2005). Lagrangian submanifolds and dynamics on Lie algebroids. ***J. Phys. A*** **38**, 24, pp. 241-308.

de León, M. and Martín de Diego, D. (1996). On the geometry of non-holonomic Lagrangian systems. ***J. Math. Phys.*** **37**, 7, pp. 3389-3414.

de León, M. and Martín de Diego, D. (1996). Symmetries and constants of the motion for singular Lagrangian systems. ***Internat. J. Theoret. Phys.*** **35**, 5, pp. 975-1011.

de León, M., Martín de Diego, D. and Santamaría-Merino, A. (2003). Tulczyjew triples and Lagrangian submanifolds in classical field theories. ***Appl. Diff. Geom. Mech.***, pp. 21–47.

de León, M., Martín de Diego, D. and Santamaría-Merino, A. (2004). Symmetries in classical field theory. ***Int. J. Geom. Methods Mod. Phys.*** **1**, 5, pp. 651-710.

de León, M., Martín de Diego, D. and Santamaría-Merino, A. (2004). Geometric integrators and nonholonomic mechanics. ***J. Math. Phys.*** **45**, 3, pp. 1042-1064.

de León, M., Martín de Diego, D., Salgado, M. and Vilariño, S. (2008). Nonholonomic constraints in k-symplectic classical field theories. ***Int. J. Geom.***

Methods Mod. Phys. **5**, 5, pp. 799-830.

de León, M., Martín de Diego, D., Salgado, M. and Vilariño, S. (2009). k-symplectic formalism on Lie algebroids. ***J. Phys. A: Math. Theor.*** **42**, p. 385209.

de León, M., McLean, M., Norris, L. K., Rey, A. M. and Salgado, M. (2002). Geometric structures in field theory. http://arxiv.org/abs/math-ph/0208036.

de León, M., Méndez, I. and Salgado, M. (1988). p-almost tangent structures. ***Rend. Circ. Mat. Palermo*** **37**, 2, pp. 282-294.

de León, M., Méndez, I. and Salgado, M. (1988). Regular p-almost cotangent structures. ***J. Korean Math. Soc.*** **25**, 2, 273-287.

de León, M., Méndez, I. and Salgado, M. (1991). Integrable p-almost tangent manifolds and tangent bundles of p^1-velocities. ***Acta Math. Hungar.*** **58**, 1-2, pp. 45-54.

de León, M., Merino, E., Oubiña, J. A. and Salgado, M. (1997). Stable almost cotangent structures. ***Boll. Un. Mat. Ital. B*** **11**, 3, pp. 509-529.

de León, M., Merino, E., Oubiña, J. A., Rodrigues, P. and Salgado, M. (1998). Hamiltonian systems on k-cosymplectic manifolds. ***J. Math. Phys.*** **39**, 2, pp. 876-893.

de León, M., Merino, E. and Salgado, M. (2001). k-cosymplectic manifolds and Lagrangian field theories. ***J. Math. Phys.*** **42**, 5, pp. 2092-2104.

de León, M. and Rodrigues, P. (1985). ***Generalized Classical Mechanics and Field Theory: A Geometrical Approach of Lagrangian and Hamiltonian Formalisms Involving Higher Order Derivatives.*** Notes on Pure Mathematics, 102. (North-Holland Publishing Co., Amsterdam).

de León, M. and Rodrigues, P. (1989). ***Methods of Differential Geometry in Analytical Mechanics.*** North-Holland Mathematics Studies, 158. (North-Holland Publishing Co., Amsterdam).

de León, M. and Vilariño, S. (2012). Lagrangian submanifolds in k-symplectic settings. ***Monatsh. Math.*** **170**, 3-4, pp. 381-404.

de León, M. and Vilariño, S. (2014). Hamilton-Jacobi theroy in k-cosymplectic field theories. ***Int. J. Geom. Methods Mod. Phys.*** **11**, 1, p. 1450007.

Libermann, P. and Marle, C. M. (1987). ***Symplectic Geometry and Analytical Mechanics.*** Mathematics and its Applications, 35. (D. Reidel Publishing Co., Dordrecht).

de Lucas, J. and Vilariño, S. (2015). k-symplectic Lie systems: theory and applications. ***J. Diff. Eqs.*** **258**, 6, pp. 2221–2255.

Mackenzie, K. (1987). **Lie Groupoids and Lie Algebroids in Differential Geometry**. London Mathematical Society Lecture Note Series, 124. (Cambridge University Press, Cambridge).

Mackenzie, K. (1995). Lie algebroids and Lie pseudoalgebras. ***Bull. London Math. Soc.*** **27**, 2, pp. 97-147.

Mangiarotti, L. and Sardanashvily, G. (1998). ***Gauge Mechanics.*** (World Scientific Publishing Co., Inc., River Edge, NJ).

Marmo, G., Saletan, E. J., Simoni, A. and Vitale, B. (1985). ***Dynamical Systems. A Differential Geometric Approach to Symmetry and***

Reduction. (Wiley-Interscience Publication, John Wiley & Sons, Ltd., Chichester).

Marrero, J. C., Román-Roy, N., Salgado, M. and Vilariño, S. (2014). Reduction of polysymplectic manifolds. To appear in ***J. Phys. A***.

Marsden, J. E. and Shkoller, S. (1999). Multisymplectic geometry, covariant Hamiltonians, and water waves. ***Math. Proc. Cambridge Philos. Soc.*** **125**, 3, pp. 553-575.

Martin, G. (1988). Dynamical structures for k-vector fields. ***Internat. J. Theoret. Phys.*** **27**, 5, pp. 571-585.

Martin, G. (1988). A Darboux theorem for multi-symplectic manifolds. ***Lett. Math. Phys.*** **16**, 2, pp. 133-138.

Martín de Diego, D. and Vilariño, S. (2010). Reduced classical field theories: k-cosymplectic formalism on Lie algebroids. ***J. Phys. A: Math. Theor.*** **43**, pp. 325204.

Martínez, E. (2001). Lagrangian mechanics on Lie algebroids. ***Acta Appl. Math.*** **67**, 3, pp. 295-320.

Martínez, E. (2001). Geometric formulation of mechanics on Lie algebroids. ***Proc. of the VIII Fall Workshop on Geometry and Physics (Spanish) (Medina del Campo, 1999)***. (R. Soc. Mat. Esp., Madrid), pp. 209-222.

Martínez, E. (2004). Classical field theory on Lie algebroids: multisymplectic formalism. `http://arxiv.org/abs/math/0411352`.

Martínez, E. (2005). Classical field theory on Lie algebroids: variational aspects. ***J. Phys. A*** **38**, 32, pp. 7145-7160.

McLean, M. and Norris, L. K. (2000). Covariant field theory on frame bundles of fibered manifolds. ***J. Math. Phys.*** **41**, 10, pp. 6808-6823.

Merino, E. (1997). ***Geometría k-simpléctica y k-cosimpléctica. Aplicaciones a las teorías clásicas de campos***. Ph.D Thesis, Publicaciones del Dpto. de Geometría y Topología, 87. Universidad de Santiago de Compostela, Santiago de Compostela, Spain.

Morimoto, A. (1969). ***Prolongations of Geometric Structures***. (Mathematical Institute, Nagoya University, Nagoya).

Morimoto, A. (1970). Liftings of some types of tensor fields and connections to tangent bundles of p^r-velocities. ***Nagoya Math. J.*** **40**, pp. 13-31.

Munteanu, F., Rey, A. M. and Salgado, M. (2004). The Günther's formalism in classical field theory: momentum map and reduction. ***J. Math. Phys.*** **45**, 5, pp. 1730-1751.

Muñoz-Lecanda, M. C., Román-Roy, N. and Yániz, F. J. (2001). Time-dependent Lagrangians invariant by a vector field. ***Lett. Math. Phys.*** **57**, 2, pp. 107-121.

Muñoz-Lecanda, M. C., Salgado, M. and Vilariño, S. (2005). Nonstandard connections in k-cosymplectic field theory. ***J. Math. Phys.*** **46**, 12, pp. 122901.

Muñoz-Lecanda, M. C., Salgado, M. and Vilariño, S. (2009) k-symplectic and k-cosymplectic Lagrangian field theories: some interesting examples and applications. ***Int. J. Geom. Methods Mod. Phys.*** **7**, 4, pp. 669–692.

Norris, L. K. (1993). Generalized symplectic geometry on the frame bundle of a manifold. Differential Geometry: Geometry in Mathematical Physics and

Related Topics (Los Angeles, CA, 1990), ***Proc. Sympos. Pure Math., 54, Part 2.*** (Amer. Math. Soc., Providence, RI), pp. 435–465.

Norris. L. K. (1994). Symplectic geometry on T^*M derived from n-symplectic geometry on LM. ***J. Geom. Phys.*** **13**, 1, pp. 51–78.

Norris. L. K. (1997). Schouten-Nijenhuis Brackets. ***J. Math. Phys.*** **38**, 5, pp. 2694–2709.

Norris. L. K. (2001). n-symplectic algebra of observables in covariant Lagrangian field theory. ***J. Math. Phys.*** **42**, 10, pp. 4827–4845.

Olver, P. J. (1986). ***Applications of Lie Groups to Differential Equations.*** Graduate Texts in Mathematics, 107. (Springer-Verlag, New York).

Olver, P. J. (2007). ***Applied Mathematics Lecture Notes.*** `http://www.math.umn.edu/~olver/appl.html`.

Paufler, C. and Römer, H. (2002). Geometry of Hamiltonian n-vector fields in multisymplectic field theory. ***J. Geom. Phys.*** **44**, 1, pp. 52-69.

Paufler, C. and Römer, H. (2002). De Donder-Weyl equations and multisymplectic geometry, in XXXIII Symposium on Mathematical Physics (Torún, 2001). ***Rep. Math. Phys.*** **49**, 2-3, pp. 325-334.

Perring, J. K. and Skyrme, T. H. R. (1962). A model unified field equation. ***Nuclear Phys.*** **31**, pp. 550-555.

Poor, W. A. (1981). ***Differential Geometric Structures.*** (McGraw-Hill Book Co., New York).

Rey, A. M., Román-Roy, N. and Salgado, M. (2005). Günther's formalism (k-symplectic formalism) in classical field theory: Skinner-Rusk approach and the evolution operator. ***J. Math. Phys.*** **46**, 5, pp. 052901.

Rey, A. M., Román-Roy, N., Salgado, M. and Vilariño, S. (2011). On the k-symplectic, k-cosymplectic and multisymplectic formalisms of classical field theories. ***J. Geom. Mech.*** **3**, 1, pp. 113-137.

Rey, A. M., Román-Roy, N., Salgado, M. and Vilariño, S. (2012). k-cosymplectic classical field theories: Tulczyjew and Skinner-Rusk formulations. ***Math. Phys. Anal. Geom.*** **15**, 2, pp. 85–119.

Román-Roy, N. (2009). Multisymplectic Lagrangian and Hamiltonian formalisms of classical field theories. ***SIGMA*** **5**, 100, 25 pp.

Román-Roy, N., Salgado, M. and Vilariño, S. (2007). Symmetries and conservation laws in the Günther k-symplectic formalism of field theory. ***Rev. Math. Phys.*** **19**, 10, pp. 1117-1147.

Román-Roy, N., Salgado, M. and Vilariño, S. (2011). On a kind of Noether symmetries and conservation laws in k-symplectic field theory. ***J. Math. Phys.*** **52**, 2, pp. 022901.

Román-Roy, N., Salgado, M. and Vilariño, S. (2013). Higher-order Noether symmetries in k-symplectic Hamiltonian field theory. ***Int. J. Geom. Methods Mod. Phys.*** **10**, 8, pp. 1360013.

Rosen, G. (1971). Hamilton-Jacobi functional theory for the integration of classical field equations. ***Internat. J. Theoret. Phys.*** **4**, pp. 281-285.

Rund, H. (1973). ***The Hamilton-Jacobi Theory in the Calculus of Variations. Its Role in Mathematics and Physics.*** Reprinted edition, with corrections. (Robert E. Krieger Publishing Co., Huntington, N.Y.).

Sardanashvily, G. (1993). ***Gauge Theory in Jet Manifolds***. (Hadronic Press, Inc., Palm Harbor).

Sardanashvily, G. (1996). Generalized Hamiltonian formalism for field theory. Constraint systems. ***Class. Quantum Grav.*** **13**, 12.

Sarlet, W. and Cantrijn, F. (1981). Higher-order Noether symmetries and constants of the motion. ***J. Phys. A*** **14**, 2, pp. 479-492.

Saunders, D. J. (1987). An Alternative Approach to the Cartan Form in Lagrangian field theories. ***J. Phys. A*** **20**, 2, pp. 339-349.

Saunders, D. J. (1987). Jet fields, Connections and Second-Order Differential Equations. ***J. Phys. A*** **20**, 11, pp. 3261-3270.

Saunders, D. J. (1989). ***The Geometry of Jet Bundles***. London Mathematical Society Lecture Note Series, 142. (Cambridge University Press, Cambridge).

Skinner, R. and Rusk, R. (1983). Generalized Hamiltonian dynamics. I. Formulation on $T^*Q \oplus TQ$. ***J. Math. Phys.*** **24**, 11, pp. 2589-2594.

Sniatycki, J. (1970). On the geometric structure of classical field theory in Lagrangian formulation. ***Proc. Cambridge Philos. Soc.*** **68**, pp. 475-484.

Szilasi, J. (2003). A setting for spray and Finsler geometry. ***Handbook of Finsler Geometry. Vol. 1, 2***. (Kluwer Acad. Publ., Dordrecht), pp. 1183-1426.

Tulczyjew, W. M. (1974). Hamiltonian systems, Lagrangian systems and the Legendre transformation in ***Symposia Mathematica, Vol. XIV*** (Convegno di Geometria Simplettica e Fisica Matematica, INDAM, Rome, 1973), (Academic Press, London) pp. 247-258.

Tulczyjew, W. M. (1976). Les sous-variétés lagrangiennes et la dynamique lagrangienne. ***C. R. Acad. Sci. Paris Sér. A-B*** **283**, 8, pp. 675-678.

Tulczyjew, W. M. (1976). Les sous-variétés lagrangiennes et la dynamique hamiltonienne. ***C. R. Acad. Sci. Paris Sér. A-B*** **283**, 1, pp. 15-18.

Vankerschaver, J. (2007). Euler-Poincaré reduction for discrete field theories. ***J. Math. Phys.*** **48**, 3, pp. 032902.

Vankerschaver, J. (2007). ***Continuous and Discrete Aspects of Classical Field Theories with Nonholonomic Constraints***. Ph.D. thesis, Ghent University. `http://hdl.handle.net/1854/LU-470403`.

Vankerschaver, J. and Cantrijn, F. (2007). Discrete Lagrangian field theories on Lie groupoids. ***J. Geom. Phys.*** **57**, no. 2, pp. 665-689.

Vankerschaver, J., Cantrijn, F., de León, M. and Martín de Diego, D. (2005). Geometric aspects of nonholonomic field theories. ***Rep. Math. Phys.*** **56**, 3, pp. 387-411.

Vankerschaver, J. and Martín de Diego, D. (2008). Symmetry aspects of nonholonomic field theories. ***J. Phys. A*** **41**, 3, pp. 035401.

Vitagliano, L. (2010). The Hamilton-Jacobi formalism for higher-order field theories. ***Int. J. Geom. Methods Mod. Phys.*** **7**, 8, pp. 1413-1436.

Weinstein, A. (1996). ***Lagrangian Mechanics and Groupoids. Mechanics Day*** (Waterloo, ON, 1992), ***Fields Inst. Commun.*** **7**, (Amer. Math. Soc., Providence, RI), pp. 207–231.

Warnick, K. F. and Russer, P. (2006). Two, three and four-dimensional electromagnetics using differential forms. ***Turk. J. Elec. Engin.*** **14**, 1, pp. 153–172.

Index